MAPPING THE NORTH

Mapping the North

MYTH, EXPLORATION, ENCOUNTER

CHARLOTTA FORSS

BODLEIAN
LIBRARY
PUBLISHING

To my parents

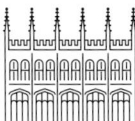

First published in 2025 by Bodleian Library Publishing
Broad Street, Oxford OX1 3BG
www.bodleianshop.co.uk

ISBN 978 1 85124 572 7

Text © Charlotta Forss, 2025

Images © Bodleian Libraries, University of Oxford, 2025, unless specified on p. 205
This edition © Bodleian Library Publishing, University of Oxford, 2025

Charlotta Forss has asserted her right to be identified as the author of this Work.

All rights reserved.

No part of this book may be reproduced, stored in a retrieval system, or transmitted in any form or by any means, electronic, mechanical, photocopying, recording, or otherwise, without the written permission of the Bodleian Library, except for the purpose of research or private study, or criticism or review.

PUBLISHER Samuel Fanous
MANAGING EDITOR Susie Foster
EDITOR Janet Phillips
PICTURE EDITOR Leanda Shrimpton
COVER DESIGN by Dot Little at the Bodleian Library
DESIGNED & TYPESET by Lucy Morton of illuminati in 10½ on 16 New Baskerville
PRINTED & BOUND by Livonia Print, Latvia, on 150 gsm Arctic Matt Art paper

British Library Catalogue in Publishing Data
A CIP record of this publication is available from the British Library

CONTENTS

PREFACE vi

INTRODUCTION
1

THE UNKNOWN NORTH
13

MAPS & FICTIONAL TRAVEL
67

ENCOUNTERS & EXPLORATION
105

ANIMALS ON NORTHERN MAPS
143

EPILOGUE 177

NOTES 181

FURTHER READING 200

ACKNOWLEDGEMENTS 203

IMAGE SOURCES 205

INDEX 209

PREFACE

This book began with an idea and a library. I became interested in how maps capture, transform and perpetuate ideas about what 'the north' is during a fellowship at the Bodleian Library in Oxford, 2018–19. The inventive ways in which map-makers who worked during widely different time periods and cultures characterized the north on maps was remarkable. Not least, I was intrigued by the amount of activity – real and fictional, human and animal – that appeared on the maps of a region that is often described as an inaccessible and barren wilderness. Thus, this book grew initially from the Bodleian Library collections and later expanded to include maps, texts and images from a range of other libraries and archives. Throughout, the fascination with the interplay between ideas about a place and the maps that portray it has remained at the heart of the study.

Beginning in classical antiquity and ending in the early twentieth century, the book is a long-term history of a concept – the north – and how it has it has manifested on maps. The purpose of this approach is both to bring forth the pervasive themes in the mapping of the northernmost parts of the world, and to show how individual maps were shaped by, and themselves took part in shaping, the idea of the north in the historical contexts in which they appeared.

I started out with a pilot study of how the north was portrayed on some of the most influential early modern European maps and atlases.

Through this analysis I identified two sets of issues that I found particularly intriguing with regard to how map-makers portrayed the north on maps: first, how southern map-makers solved the problem of mapping a region that they knew very little about; second, how they populated and depopulated this unknown region with humans and animals. The first set of issues draws attention to how knowledge about the geography of the north was constructed and made credible in different historical contexts. The second set lays bare the economic and social forces that underpin all maps, and it reveals the power structures and real-life consequences that resulted from mapping. I expanded the investigation by tracing these two sets of issues backwards and forwards in time, paying attention to what changed or remained the same between historical contexts. The first two chapters are devoted to the issue of constructing a credible map of the geography of the north. The second two chapters deal with the roles of people and animals in the mapping of the north.

    The chapters of the book, as well as the individual maps, images and texts discussed, shift between different viewpoints and conceptions of what, and where, the north is and where it was believed to be. To this end, discussions about well-known maps complement lesser-known examples. Throughout, I have contextualized the maps by analysing them together with and in relation to contemporary discourses expressed through images and texts. I have also striven to include a diversity of maps that showcase cartographic richness and variety to highlight the persistent themes that have constituted the evolving notion of the north. Nevertheless, more European travellers and map-makers than people from other parts of the world appear on these pages. This focus is a result of my own field of expertise: northern Europe in the early modern period (*c.* 1450–1750). It is my hope that others will widen this exploration and take up where I leave off.

INTRODUCTION

THE LARGE SIXTEENTH-CENTURY wall map *Carta Marina* makes a claim about what and where the north is.[1] The full title of the map states that this is a 'Marine map and description of the northern lands and the wonderous things contained in them'[2] (*Carta Marina et descriptio septemtrionalium terrarum: ac mirabilium rerum in eis contentarum*). The map outlines the contours of land and sea in northern Europe, with Iceland and a partial outline of Scotland in the west, Scandinavia and the coastlines of the Baltic Sea at the centre of the map and western Russia in the east. The map represented, at the time of its making, a leap in geographical detail that greatly interested its European audience (fig. 1).

This showcasing of the north on *Carta Marina* was by no means coincidental. The map was published in 1539, amid the religious turmoil of the Reformation. Its maker, the sixteenth-century Swedish scholar and theologian Olaus Magnus, intended the map to be a forceful reminder to papal Europe that the Nordic countries had value and potential, hoping that this realization would lead to efforts to reverse

---

1 OPPOSITE & PRECEDING SPREAD *Carta Marina* (1539, 1572) is a large map, 125 × 170 cm. Olaus Magnus, its maker, also had large ambitions. In the bottom right corner, the image of a lion and a mouse references Aesop's fable about a mouse saving a lion, reminding papal Rome of the worth of the distant Nordic countries.

the Reformation in the homeland he had left as a religious exile. With *Carta Marina*, Olaus Magnus wished to emphasize the wondrous nature and culture of the north and illustrate what Rome had lost by Sweden's and Denmark–Norway's change of religious allegiance.

Carta Marina did not help Olaus Magnus succeed with his ambition to reverse the Reformation in the Nordic countries. The map nevertheless became highly influential in spreading information about the northernmost parts of Europe.[3] On *Carta Marina* the hitherto relatively unknown region east across the North Sea and north across the Baltic Sea was filled in with new and tantalizing knowledge about geography and ethnography. This knowledge about the north was, though, far from neutral. While the map contains information about the shape and relative position of islands, coastlines and mountain ranges, Olaus Magnus imbued the land with characteristics that spoke to contemporary conceptions of what the north was, and he emphasized features which were calculated to help him in his cause, such as rich natural resources, long and noble history, and exotic cultural habits. Thus, through a combination of reiteration and reinvention, *Carta Marina* presented geographical knowledge infused with religious, cultural and political ideologies.

Throughout history, innumerable maps have been made which, like *Carta Marina*, claim to show their viewers 'the north'. These maps portray widely varying places, they were made for diverse purposes, and they were used for an even broader set of objectives. Many were less famous and less influential than *Carta Marina*, yet all reflect a continuous conceptualization over time of the idea of the north.[4] When Olaus Magnus outlined Sweden and Denmark–Norway on his map and called them 'the northern lands', he defined what the north was and what it was not, rather than simply depicting an already existing entity.

At the same time, maps say as much about the priorities of the situations and places where they were made, as they give information about the geography of the northernmost parts of the world. The key to understanding *Carta Marina* is thus located as much in Italy – where the map was published, where Olaus Magnus lived, and where he saw his primary readership – as in Scandinavia.[5]

This book follows the story of how maps of the northernmost parts of the world, over the course of centuries, took part in an ongoing making of the north as an idea, while simultaneously they were made and used to address specific issues in their own times. In the process, we encounter hardened Arctic explorers keen to promote their own legacy as adventurous travellers, and we meet Inuit informers depicting their home environs in an often exploitative knowledge trade. We encounter map-makers who viewed the north from far away, projecting the concerns and priorities of Europe into the Arctic, and Nordic statesmen and scholars who tried to reclaim the identity of the north to make it their own. We encounter humans living and travelling through the lands and seas of the circumpolar north, and we meet the animals that they studied, hunted and exoticized.

MAPS AND THE IDEA OF THE NORTH

Among scholars interested in the history of mapping, considerable attention has been devoted to the history of maps depicting the northernmost parts of the world.[6] However, it is only in recent years that researchers have begun to address the implications of map-making for how the north as a concept has been imagined and understood.[7] In fact, mapping endeavours and ideas about the north are processes that, through much of human history, have been closely bound together. For example, the expeditions that resulted in increased knowledge of, and ideas about, the northernmost parts of the world have frequently been discussed in terms of 'improving the map' of the north, and the results of travel in the north have been conveyed in part through maps. In the words of Robert E. Peary, who claimed he was the first explorer to reach the North Pole, his expedition of 1909 had filled another blank stretch of 'the world's map'.[8]

This book is not a historical survey of the making of the northern part of the 'world's map', yet it touches on many of the themes that would be central in such a survey – the harsh climate, the cross-cultural encounters and exploitations, the political and economic incentives behind mapping – and uses examples of individual maps to explore what the drive to map the circumpolar north meant for perceptions about this part of the world at different points in time. The spatial nature of maps ties knowledge and ideas to places, real or imagined. Maps of the north are crucial because they have been important historically in the making of knowledge about the north, and in the process took part in shaping that knowledge.

What, then, does a study of maps specifically contribute to our understanding of the idea of the north? To begin with, investigating mapping and the idea of the north jointly draws attention to how weather and climate set the terms for expanding our knowledge concerning the northernmost parts of the world. One of the main reasons why certain places remained blank spots on maps was their inaccessibility to the rest of the world, and that in turn was often a result of their geography and climate. While this can be seen around the world, the circumpolar north arguably takes first prize among hard-to-map places.[9] The history of the map of the far north is filled with the attempts and failures of explorers, and this was in large part due to difficult climatic conditions.

As a result, the north remained for a long time a partly mythical place to outsiders; it was a place where real and imagined geography could exist side by side. Literary scholar Alfred Hiatt calls this phenomenon of places that were imagined long before they were visited 'conjectured land'.[10] As map historian Chet Van Duzer has noted, it was often such regions far away and out of reach that early map-makers tended to populate with strange and wondrous creatures.[11] While many of the areas around the world which European map-makers had believed held supernatural wonders turned out to be lacking in such, the potential of the north as a repository for future discoveries of the strange and wondrous was pervasive because it remained more difficult to access. Instead, people have continually speculated about what the far north looked like and who might live there, and one of the ways in which such speculation has been articulated is through maps. This has resulted in a range of particularly interesting and creative maps – some of which will be familiar to readers, and others of which will have received relatively little attention to date.

With the stark realities of global warming, the unique characteristics of the circumpolar north are changing.[12] The north is becoming considerably less cold and, with technological advances, less inaccessible. Still, the idea of a north far away from the concerns of the rest of the world is arguably no less pervasive today. Laying bare the history of this idea is one step towards dismantling the fallacy of an isolated north. History helps us create perspectives on the ongoing transformations of the natural world, and how they link with political and cultural ideas about the characteristics of the north as a place.

Another reason to pair mapping and the idea of the north, and to follow this issue over the course of centuries, is that the map medium nuances the cold and desolate north that abounds in historical as well as present-day accounts of the region. Literary scholar Jen Hill has noted that for Britain in the nineteenth century the exploration and mapping of the north was construed as happening in an empty space, and that this spoke to contemporary masculine ideals of national identity.[13] With a longer time perspective, the maps discussed in this book show how mapping of the northernmost parts of the world resonated with, but also at times diverged from, such ideas about the north as a far-away, cold and

2 OVERLEAF The early modern Swedish state harboured ambitions well beyond its borders. This 1626 map by Andreas Bureus includes Sweden's neighbouring countries and its title, *Orbis Arctoi* ('The Arctic World'), references a broader region. The cartouche of King Gustavus Adolphus, 'The lion from the north', and his wife Maria Eleonora presents the 'Arctic World' as Swedish.

DELINEATIO Auctore Andrea Bureo Sueco

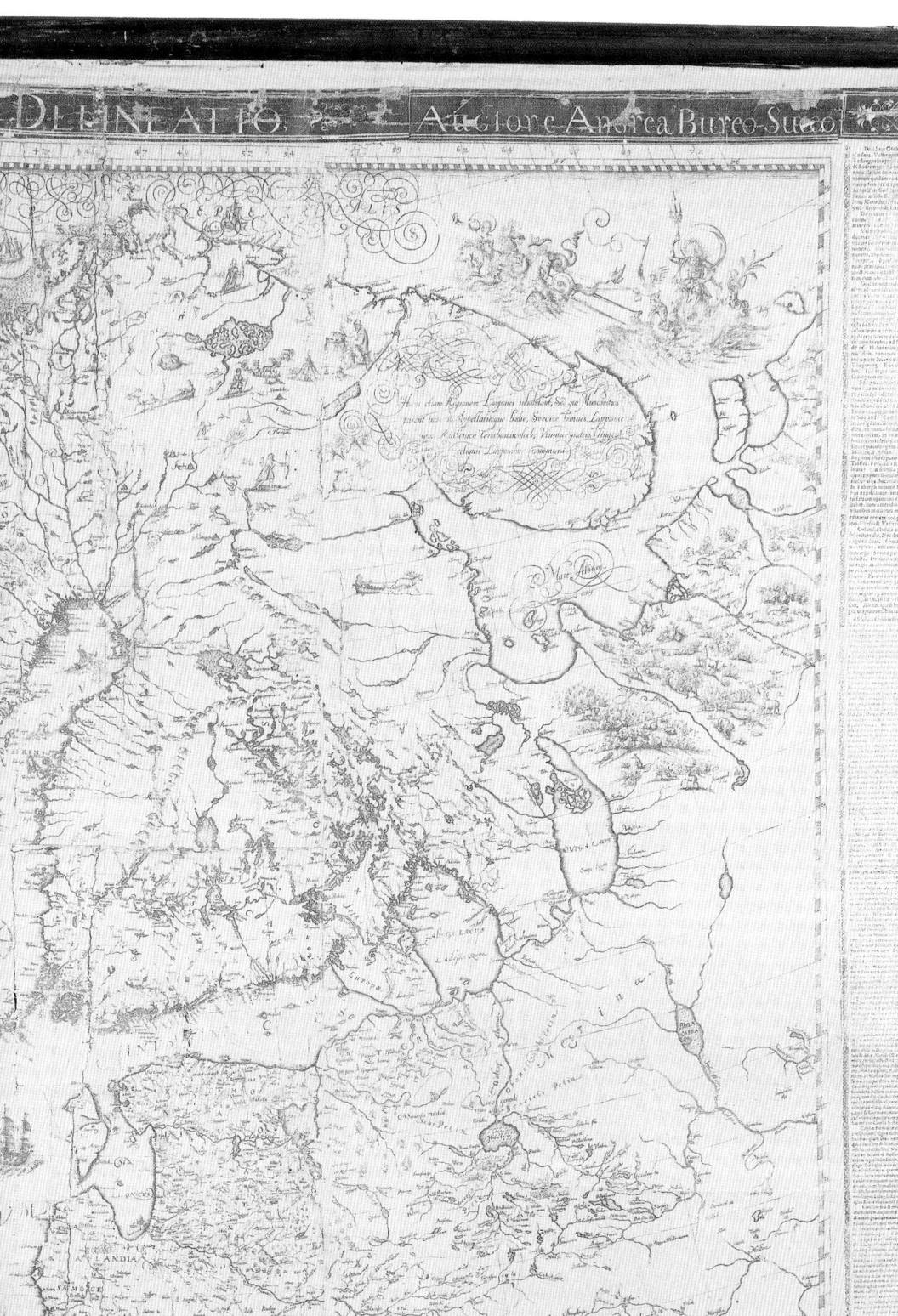

inaccessible place. As we will see, maps portraying ice and emptiness have been common over a long period of time. At the same time, many maps of the north are filled with activity, human as well as animal. Focusing on the roles of these people and animals in the mapping process grounds the discussion about the idea of an empty north in the economic and social contexts that were crucial in the making of the maps.

This is not to say that maps have been the sole medium to shape perceptions of the northernmost regions of the world. On the contrary, ideas about the north reach us through genres such as news reports, fiction and travel writing. Many of the most interesting features of maps appear precisely when they are discussed together with the texts and images with which they were made and used. Hence, while the focus of this book is on maps, they are discussed within the historical contexts in which they appeared.

PLACING THE NORTH ON A MAP

There is no one way of defining the north. How map-makers have gone about placing and defining the north on maps has been a question of artistic preference or chance, but also the result of genre conventions, resources and ideology. To geographers in Ancient Greece the north was a mythical place beyond the mist-clouded ocean. It was a land of monsters and half-believed tales. In contrast, to seventeenth-century Swedish statesmen and scholars the north was 'us'. In a highly political attempt to present their own country as prominent in geopolitical, historical and cultural terms, they referred to the north as a placeholder for Sweden. In cold-war politics, the Soviet Union was described as the east; yet in the Renaissance, Russia was considered to be located in the north. Nor is the location of the north stable within states. Scholars note that the opinion as to where 'north' is in Canada has changed northward over the last century.[14]

Indeed, in one interpretation the north is everything to the north of where you are standing at the moment.[15] When talking about the

3 *HMS Erebus in the Ice* by François Étienne Musin, 1846. HMS *Erebus* is dwarfed by looming icebergs and its crew is seen struggling against the forces of nature. The fate of Sir John Franklin's 1845 expedition caught the imagination of European and American audiences and became an impetus for map-making.

northernmost reaches of the world, scientists as well as politicians today often use the Arctic Circle as a demarcation line, encompassing all land and sea located above 66.5°N in the Arctic region. Another option is to use climate, defining the Arctic – a term at times used as a synonym for the north – as the area which has an average monthly mean temperature that does not rise above 10°C during the warmest month of the year. With the stark realities of climate change, a smaller region can be called Arctic today than in the past.

Politics are never far removed from geography, and the north is often defined along the lines of the states that have territory and strong interests in or close to the Arctic Circle. For example, the collaborative

INTRODUCTION 9

intergovernmental organization called the Arctic Council consists of the eight countries that have part of their territory above the Arctic circle: Canada, Finland, Greenland (Denmark), Iceland, Norway, Russia, Sweden and the United States, as well as six organizations representing indigenous peoples: Aleut International Association, Arctic Athabaskan Council, Gwich'in Council International, Inuit Circumpolar Council, Russian Association of Indigenous Peoples of the North and the Sami Council.

To complicate matters further, north is not only a cardinal direction but also a location and a set of ideas. Historians have studied the contours of how people have associated 'the north' as a concept with certain characteristics, emphasizing in particular how it has been understood as a wild and inaccessible place, a place of bounty but also of danger. And, more than anything else, the north has been conceptualized as a place of cold and ice.[16] These ideas about the north are cultural in nature, beyond their correspondence to any measurable temperatures or frozen seas. At various points in time, the idea of a cold and faraway north has also been used for economic and political gain, to define 'us' and 'them', to moralize and to extoll the virtues of individuals or societies. Nor is the idea of a cold north a thing of the past, despite the changing climate of our present day. A vaguely defined north continues to be understood in popular culture today in terms of a cold, unspoiled and potentially wild natural environment.

Often, there is more which separates than unites the north on different maps. Still, trends that otherwise remain obscure come to light through contrasting and comparing over time; simultaneously, features that appear pervasive when observed in a given context can dissipate in the light of comparison. How have individual map-makers solved the conundrum of placing the north on a map? On what grounds have map-makers delineated the north as a place? Who gets to decide where the north is, and how has this changed over time and between cultural and political contexts? What has been considered credible knowledge to include on a map, and what has been left out? The present book sets out to begin to answer these questions.

The chronological starting point of the book is antiquity and the mythical islands that early Mediterranean travellers and geographers located in the frozen seas of the North Atlantic Ocean. The themes about the north established in antiquity remained influential well into the modern era. The main focus of the investigation is on how medieval and early modern maps portrayed the north, and how these portrayals changed in the mapping of the nineteenth and early twentieth centuries. The end point of the analysis is the first expeditions that reached the North Pole – or at least claimed that they did – and the maps made in relation to explorations in the early twentieth century. To some extent this is an arbitrary end point. After all, we still harbour ideas about the north in the present day. Still, the relation between maps and knowledge with regard to the north has changed with the completion of Peary's so-called 'world's map', and these transformations and the ongoing conceptualization of the north into the modern era warrant their own separate study.

The primary emphasis of the book is European maps, though I have striven to put European views in perspective through highlighting map-makers who are indigenous to the north, as well as providing discussion of and comparison to other viewpoints, such as medieval Arab conceptions of the north. I have selected the maps discussed in this book to include well-known and influential works, but I have also attempted to show the variety of ways in which people have imagined the north through maps. The approach used cannot capture all nuances of meaning in how the north as a concept has been used on maps; nor can it do justice to the innumerable map-makers who have depicted the north. Rather, this wide range of material is chosen to trace conceptual patterns and probe how themes in the mapping of the north have persisted and changed over time.

THE UNKNOWN NORTH

For a very long time the northernmost reaches of the Earth were a known unknown: a region known to exist given the spherical shape of the Earth, yet beyond the direct experience of people living far away and further south.[1] When making a map that purported to show the world in its entirety or a map of the northernmost parts of the world, map-makers were faced with the difficulty of depicting a region that was – at different points in time – completely or at least partially unknown. They went about solving this paradox of the northern known unknown in different ways, and in the process ideas about what the north was took form.

More than 2,000 years ago Greek- and Latin-speaking scholars began to discuss what the northern limits of the inhabitable Earth looked like, describing it as an inaccessible place with strange natural phenomena. Early Jewish scholars built on this template, proposing that signs about the forthcoming end of the world would first appear from the north, thus giving this far-away north a decidedly sinister cast. This theme was picked up by medieval Christian and Islamic writers and map-makers. In their maps the view of the north was shaped and

4 OVERLEAF Scholars in antiquity located the farthest north of the inhabitable Earth at 63°N and the fictional island Thule, reached when sailing north from the British Isles. Further to the east in Asia, Scythia was believed to extend towards the north, as shown on this Ptolemaic world map.

reshaped in relation to religiously informed ideas about the world and our place in it. During the Renaissance, European map-makers combined old accounts with new discoveries to create maps of the north that would be reliable for a readership hungry for geographical wonders.

The depiction of cold and harsh weather conditions has long been a common feature of maps of the north, emphasizing the climate as an essential, and essentially exotic, feature of this part of the world. This idea of the cold, dangerous north caught the imagination of a broader audience during the nineteenth century as explorers ventured ever further north. The maps of the far north both reflected expansion in knowledge about the geography of the north and presented this knowledge as precariously incomplete. Paradoxically, from the fifteenth century onwards, the fact that the far north was known to be unknown became integrated into an account of the growth of geographical knowledge. Throughout this long history, the north functioned as a repository for projecting religious and political concerns that were important in the map-makers' own respective societies.

ULTIMA THULE

Geographers and astronomers in Greco-Roman antiquity were aware that the Earth was round. In the third century BCE the Alexandrian scholar and librarian Eratosthenes calculated the Earth's circumference and reached a figure that was close to what we now know the Earth to be – an impressive feat before the age of satellites and other precise instruments of measurement.[2] However, ancient geographers did not know much about what the far north looked like and they believed that large parts of it were uninhabitable. A map made following the instructions of the famous second-century geographer Claudius Ptolemy ends well below the North Pole with the mythical island Thule at 63°N, representing the northern bounds of geographical knowledge (FIG. 4).[3] As a reference, present-day locales at 63°N include St Lawrence Island in the Bering Sea, the northern part of Hudson Bay, and Surtsey, the southernmost island of Iceland.[4] The Russian Siberian city of Yakutsk is located just south of 63°N.

Somewhat ironically, it is not entirely clear if Claudius Ptolemy made any maps, because none made by him has been preserved. Still, his works on geography and astronomy include discussions about maps and instructions on how to make maps based on distances between different locations. These instructions created a framework for generations of map-makers for how they thought about mapping. Thus, a map made by following Ptolemy's instructions gives some approximation to how Ptolemy himself would have imagined the world.

Ptolemy's map is usually called 'world map', though it only claims to show the known inhabited world, the *oecumene*. In the west, Europe and the coast of 'Lybia', or Africa, is bounded by the world ocean, which the Greeks thought encircled the world. China and the Malaysian peninsula constitute the easternmost part of the map. The Indian Ocean is portrayed as an inland sea bordered to the south by 'unknown land'. In the north, the island Thule is situated just north of the British Isles and next to the contours of land representing Scandinavia. Further east, a large land mass is identified as the 'unknown' Scythia. All of the more detailed descriptions of places and coastlines are found in the narrower confines of the Mediterranean Sea and the Black Sea.

It is not surprising that the geographical knowledge of Eratosthenes, Ptolemy and their contemporaries focused on the Mediterranean Sea, as that was the region where they lived and around which politics, science and culture revolved. Still, ancient geographers were interested in what the world looked like further away. Ptolemy emphasized that his subject matter was the description of specific places as well as the description of the entirety of the inhabited world, even if he knew little about some of its parts.[5]

The island Thule, which marked the northern bounds of geographical knowledge in Ptolemy's world, was first mentioned by Pytheas of

Massalia (Marseille), who sailed through parts of north-western Europe around the year 325 BCE. Pytheas' travel account, poetically named *On the Ocean*, is lost to history, but it was discussed by numerous Greek and Roman authors whose writings have been preserved. Interestingly, there were different opinions on the veracity of Pytheas' account already in antiquity, and scholars still debate the identification of Thule in the real world. While early modern map-makers often identified Thule with Iceland, later commentators argued that Pytheas was referring to Norway or the Shetland Islands.[6]

What we know of Thule from ancient geographers is that it was an island which could be reached after six days of sailing from the British Isles, that it was near the frozen sea, and that during parts

5 Europe is represented as an island on this Islamic world map. In the lower left-hand corner, the wall of Gog and Magog indicates a northern region beyond the day-to-day experience of the Egyptian map-maker.

18   MAPPING THE NORTH

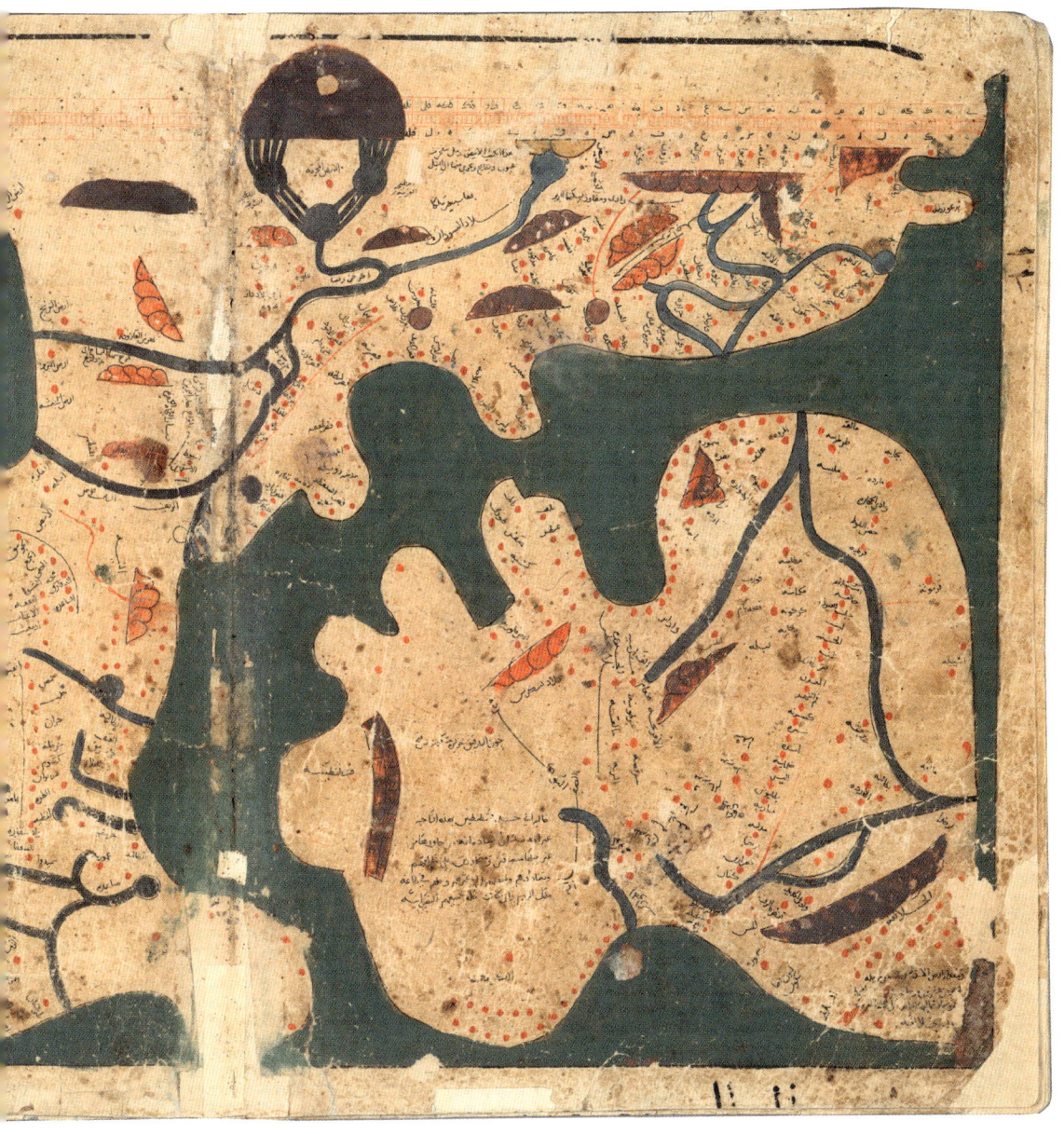

of the year 'the night becomes very short, 2 hours for some, 3 for others, so that, a little while after setting, the Sun rises straightaway'.[7] Regardless of whether Pytheas of Massalia actually visited the place he named Thule or relied on hearsay, this characterization will probably seem familiar to most people who have visited north-western Europe, where Thule was supposedly located, and experienced the short summer nights. That probably explains why generations of writers and map-makers began to include the island of Thule in descriptions of the north, even though the name itself was more a myth than an actual place in the world. Over time, the expression *Ultima Thule* ('furthest Thule') began to take on a less literal meaning, not only describing the location of Thule but also signifying that it was far away from the central Mediterranean and on the edge of the inhabitable world.[8]

MEDIEVAL ARAB, JEWISH AND CHRISTIAN PORTRAYALS OF THE NORTH

Early Islamic astronomers and geographers built on the knowledge of Greco-Roman authors, and not least on Ptolemy's advice about how to make maps of the world and its various parts. Like earlier geographers, early Islamic map-makers had little first-hand knowledge about the northernmost parts of the Earth. Still, a combination of information and conjecture about far-away places permeated their maps.

These trends can be seen in a beautiful manuscript treatise entitled *Book of Curiosities of the Sciences and Marvels for the Eyes* (in Arabic *Kitāb Gharā'ib al-funūn wa-mulaḥ al-ʿuyūn*), housed in the Bodleian Library, Oxford (FIG. 5). The *Book of Curiosities* is an anonymous cosmographical treatise, including a set of rare and impressive maps, composed in Egypt in the eleventh century (between 1020 and 1050).[9] One of the maps in the volume is an unusual rectangular world map.[10] Like other early Islamic world maps, it is oriented with south at the top. The south orientation makes the map visually different from medieval European world maps, commonly oriented with east at the top of the map, or Ptolemaic world maps produced in early modern Europe, which were oriented with north at the top. Still, Ptolemy was an important source for the map and for the accompanying account of geography. In fact, the section of the treatise that describes the geography of the world opens with the words 'Ptolemy said'.[11] With regard to the northernmost reaches of the globe, the *Book of Curiosities*, like many other medieval Islamic geographical works, took note of places like the mythical island Thule, the knowledge of which came from Greco-Roman authors.[12]

Over time, Ptolemaic knowledge became less relevant for Islamic map-makers. Historian Yossef Rapoport points out that Islamic map-makers, such as the maker of the world map in the *Book of Curiosities*, considered mathematical precision in their maps less important than, for example, leaving room for textual labels and instructions to users. This made Ptolemy's mathematical models for map-making superfluous, and, as a result, by stages Ptolemy became an outdated authority.[13]

For more immediate knowledge about the northernmost reaches of the world, there were reports by Islamic travellers who had encountered people living in the faraway north. One such example is the account of Aḥmad Ibn Faḍlān, an Arab Muslim who, in 921, travelled along the Volga as part of a mission from the caliph in Baghdad. Ibn Faḍlān described his encounter with a people called the 'Rūsiyyah'. Historians and archaeologists debate whether the Rūsiyyah were in actuality Scandinavian Norsemen, Slavs, or perhaps a mixture of several cultures. Regardless of the exact identity of the Rūsiyyah, Ibn Faḍlān's account gives insight into how an Arab traveller in the tenth century interpreted people who were very different from himself.[14] One of the aspects Ibn Faḍlān emphasized was that the Rūsiyyah were unclean. He noted that they spat in the water where they washed their hands. To a Muslim, ritual cleansing was an important religious ritual, and Ibn Faḍlān saw it as a marker of moral decay that the Rūsiyyah did not attend to personal hygiene in a manner that was familiar to him. In the account, the passage served the purpose of emphasizing the strangeness of the northerners.

The rectangular world map in the *Book of Curiosities* also reflects the concerns and priorities of home. The map centres on Fatamid Egypt, where it was made, and the Arabian Peninsula – with Mecca marked by a yellow half-circle – takes up a significant part of the left-hand side of the map. On the right-hand side the Mediterranean Sea hugs Europe, which is portrayed as an island. The places marked on Europe-the-island refer mostly to Spanish and Italian localities, and the centrality of the Mediterranean Sea is emphasized again in other maps in the volume. The connections to northern and eastern trading networks, like those travelled by Ibn Faḍlān, are seen in a reference to Kyiv in present-day Ukraine, which is located in the lower left-hand corner of Europe. However, the northernmost parts of the map were not the focus of its maker. Like Ptolemy, the map-maker provided more information and gave more prominence to places closer to home.[15]

A large wall and a gate occupy the lower left-hand corner of the map (north-west on the map). This wall corresponds to a constructed barrier that early Jewish, Christian and Islamic thinkers believed protected the world from Gog and Magog, the ravishing hordes of the north.[16] Gog and Magog were part of ancient and medieval religious visions of the coming of the end of time, but they were also frequently discussed as real and identifiable people living in the far north. It is both in their capacity as portents of the coming apocalypse and as potentially real peoples that Gog and Magog appeared on medieval Middle Eastern and European maps. This makes them a fascinating example of how geographical and religious thinking could intertwine, and point to the joint cultural heritage of Judaism, Christianity and Islam.

6 In Islamic tradition, Iskandar – Alexander the Great – became associated with the legend of Gog and Magog. In this sixteenth-century Persian painting, Iskandar is assisted by a group of humans and demons when constructing a wall to protect civilization against the hordes of the north.

The Jewish prophet Ezekiel, who lived during the Babylonian exile in the sixth century BCE, described how Gog and his peoples would break free from the north, 'coming like a storm, covering the land like a cloud'.[17] The invaders would destroy homes and livelihoods, and in the process signal that the end of time had come. In accounts like these, Gog and Magog were seen as harbingers of the apocalypse. Their appearance on a map, like the world map in *Book of Curiosities*, would work as a reminder of imminent cataclysm.

Early Christians elaborated on the idea of a threat in the north. Gog and Magog sometimes referred to entire peoples, sometimes to the regions where these people lived, and sometimes to the princes that would lead the coming onslaught. Over time, other traditions were interwoven with the idea of Gog and Magog. Not least, Alexander the Great – the famous ruler of Macedonia (356–323 BCE), who conquered large parts of the Middle East and the Mediterranean region – was believed to have built an enclosure to keep Gog and Magog at bay. A Syriac poem from the seventh century describes in some detail how Alexander built a gate to fill in the gap between two mountains, thereby barring the way of the dangerous northern people:

> King Alexander made haste and made the door against the North, and against the [robbers] and the offspring of Magog.[18]

The legend of Alexander the Great saving humanity from Gog and Magog emerged both from a broader fascination with Alexander the Great and from the expectation that existence as we know it will end when the nightmares of the apocalypse are released on the world. Historians Emeri van Donzel and Andrea Schmidt also note that many early Christian accounts of Gog and Magog echo invasions that people

7 This small *mappa mundi* (170 × 130 mm) was included in a psalter made in mid-thirteenth-century England. The map literally has a religious framing, with a Christ figure embracing the world. In the north-east (top left of the map), Alexander's walls protect humanity from Gog and Magog.

THE UNKNOWN NORTH

experienced in their lifetime; for example, the attacks by nomadic tribes on the Roman Empire in the fourth through seventh centuries.[19]

In the Koran, Gog and Magog are described as a menace, kept at bay by the barrier built by 'the two horned one' – a title also used as a reference to Alexander the Great. A rich tradition developed, where Gog and Magog were described as semi-monstrous peoples, ready to invade the civilized world. Not least, geographical works, travel accounts and maps mentioned Gog and Magog, as seen on the rectangular world map in the *Book of Curiosities*.[20] The actual place that these authors and map-makers argued was the home of Gog and Magog varied, but it was always located just beyond the lands they themselves knew well. As so often, lands far away offered an opportunity to speculate about the possible existence of monstrous and threatening people.

The image of Gog and Magog ensconced behind their wall also appears in medieval Christian world maps, or *mappae mundi*. These circular or oval maps portray a religiously informed world geography.[21] On a *mappa mundi* the world is shown as a disc surrounded by the world ocean. The continents are separated from each other by the centrally placed Mediterranean Sea, the Red Sea and the river systems of present-day Russia. Asia can usually be found at the top of the map, with Europe (left) and Africa (right) taking up the lower part of the map. A *mappa mundi* reflects knowledge about places in the physical world; yet it also had symbolical meaning. To a medieval scholar and map-maker it would not have made sense to separate a description of world geography from religious matters. As God's creation, the world was inherently religious.

8 World map appearing in a Bodelian Library copy of Ranulf Higden's *Polychronicon*. Toponyms have been added to the north-western part of the world ocean (lower left-hand corner of the map), including England, Ireland, Iceland, Thule and the enigmatic 'Wyntlandia', possibly referring to Finland or Vinland.

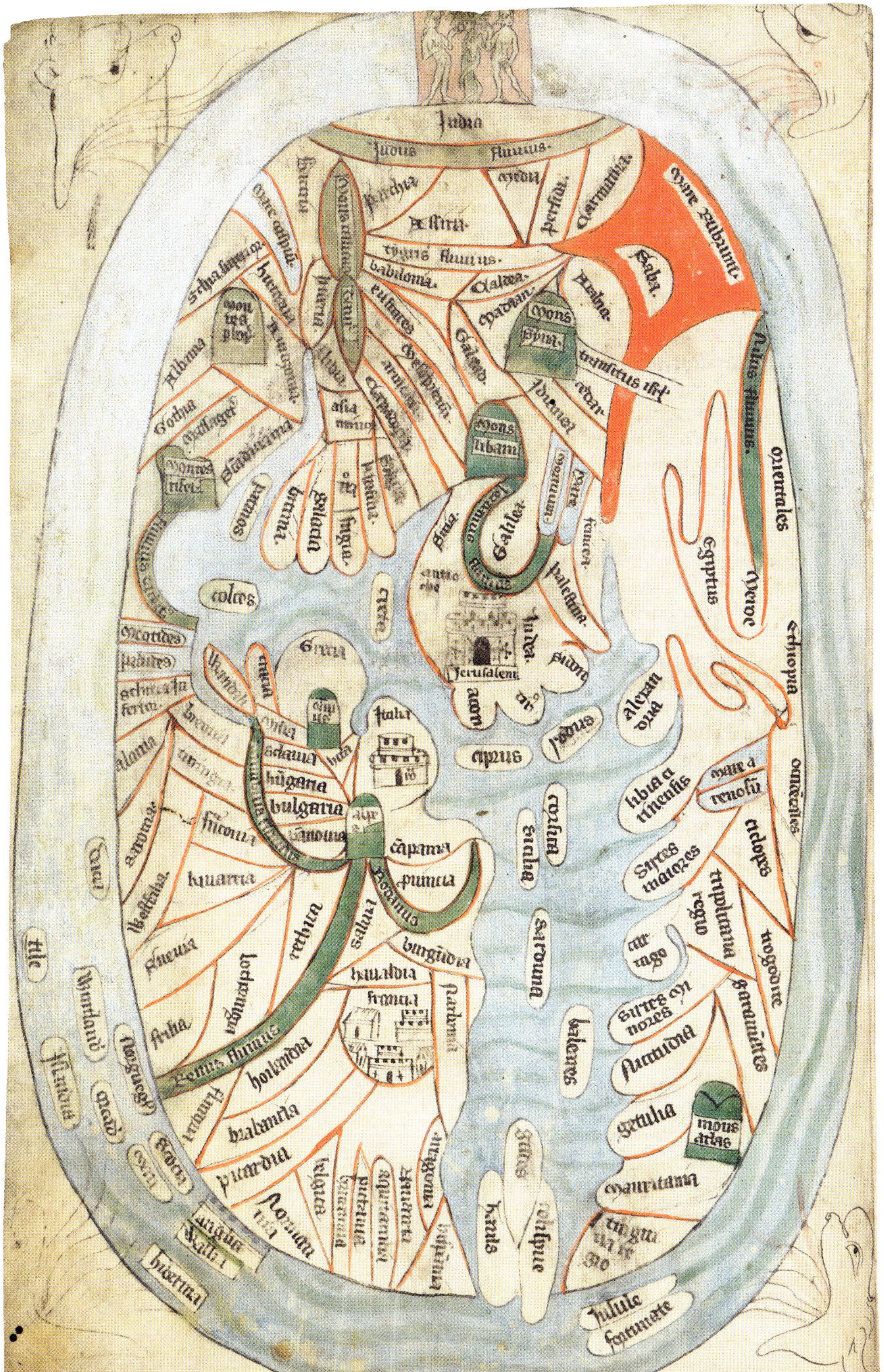

Judea
Judus fluvius
Media
Persida
Carmania
Mare rubrum
Saba
Arabia
Assiria
Tygris fluvius
Babilonia
Caldea
Madian
Mons Synai
Nilus fluvius
Mons libani
Galilea
Palestina
Judea
Jerusalem
Egyptus
Meroe
Ethiopia
Orientales
Alexandria
Colcos
Grecia
Creta
Italia
Cyprus
Rodus
Sicilia
Sardinia
Baleares
Bulgaria
Saxonia
Bavaria
Campania
Francia
Burgundia
Narbonia
Holandia
Brabantia
Picardia
Normannia
Aquitania
Hispania
Renus fluvius
Mauritania
Getulia
Numidia
Mons atlas
Gades
Insule fortunate
Hibernia
Anglia
Tile
Norguegia
Albania
Gothia
Massagete

CAVRVS·CHORVS·VEL·IAPIX·SIVE·ARGESTES
FAVONIVS·ZEPHIRVS
mare glaciale
EVROPA
DATIA
Pontus euxinus
LIBIA INTERIOR
AFFRICA
Circulus equinoctialis
ETHIOPIA·INTERIOR
AFRICVS·VEL·LIBS
LIBONOTVS·EVROAVSTER

AQVILO VEL BOREAS
CECIAS APELIOTES
SVBSOLANVS
ASIA
Sacharũ regio
Serica regio
Scitia extra imaum montẽ
Emodis mons
Tropicus cancri
India extra gangẽ fluuium
India intra gangẽ fluuium
Sinus magnus
Sinus gangeticus
MARE INDICVM
MARE INDICVM
PRASODVM
MARE
Terra incognita secundum ptholomeum
Gradus longitudinis ab occidente in orientem
Latitudo
WLTVRNVS EVRVS
EVRNOTVS

This religious framework is prominent in the *mappa mundi* that accompanies the English Benedictine monk Ranulf Higden's world-historical treatise *Polychronicon*, written in the fourteenth century. Higden's text is a detailed account of chronology and geography, and as such it gives an insight into what knowledge learned Christians in medieval Europe considered important. The maps that accompany several of the surviving manuscripts give added spatial emphasis to the world-view presented in the text. One of these maps is preserved in the Bodleian Library in Oxford (FIG. 8). Looking at the map, the importance of religious world-views is seen, for example, in the over-sized representation of Jerusalem close to the centre of the map. At the top of the map (in east Asia) the map-maker has included a depiction of Adam and Eve in Paradise. Yet other legends on the map take note of geographical locations, like Ireland, Italy and the 'Insule Fortunate' (the Canary Islands). As was the case with other medieval European maps, a blend of places in the map-maker's contemporary world and religiously significant themes are portrayed next to each other.[22]

PTOLEMY 2.0

In the early fifteenth century, European scholars again began to read, translate and reproduce Ptolemy's work on geography.[23] Not least, Renaissance scholars soon started following Ptolemy's instructions on how to make maps (as seen in FIG. 4). Unlike their Islamic counterparts, early modern European map-makers oriented their Ptolemaic world maps with north at the top of the map.[24] At first they were content to reproduce the geographical coordinates of the approximately 8,000 places that Ptolemy provided, but they soon found a need to amend and improve Ptolemy's maps to fit the constantly expanding world of their

9 PREVIOUS SPREAD In this fifteenth-century version of Ptolemy's world map, the frames of the map have been expanded in the north to include new geographical information that was unknown to ancient geographers. Legends emphasize that this is a frozen region.

own age. Renaissance scholars began updating Ptolemy's maps with information that was reaching them through reports from explorers, as well as from older sources that were discovered anew. As a result, the geographical boundaries of Ptolemy's maps were – literally – breached.

This can be seen in one of the first printed editions of Ptolemy's *Geographia*, published in the German city of Ulm in 1482 by the printer Leinhart Holl and with maps by the Benedictine monk Nicolaus Germanus. The layout and projection of the Ulm world map follows Ptolemy's instructions closely, but in the north-western corner of Europe the geography seems to have taken over the map's scale bar, interrupting the specification of longitudes with additional geographical information (FIG. 9). 'The frozen sea' (Lat. *mare glaciale*) extends beyond the original boundaries of the map, and a large peninsula stretches north. This version of Ptolemy's map did not so much contradict ancient geographical knowledge as update it. After all, Ptolemy had set out to describe not the whole world, but the world known to him.

In addition to updating Ptolemy's existing maps, Renaissance scholars made new maps using Ptolemy's methods. Some of the earliest maps added to collections of Ptolemy were maps that depicted Scandinavia. Thus the oldest surviving map of Scandinavia was made by the Danish scholar Claudius Clavus around 1427; it forms part of a manuscript copy of Ptolemy's *Geographia* housed in Nancy, France.[25] Clavus used Ptolemy's system of coordinates to make his map, but he also drew on his own knowledge of local geography, as well as on medieval scholarly works on Scandinavia, like Adam of Bremen's history of Christian missions to the north from the archbishopric of Hamburg–Bremen, and the twelfth-century Danish historian Saxo Grammaticus's *Gesta Danorum*. Clavus presented a somewhat compressed version of

10 OVERLEAF The Danish scholar Claudius Clavus used his local experience in combination with textual sources to create the oldest surviving stand-alone map of Scandinavia. Iceland is shown as a semicircular island in the vividly coloured sea. The contours of 'The province of Greenland' are visible to the west.

Europa ta

Claudius Ptholomeus

Cerebrum infidelium regio maxie septetrional · Unipedes maritimi

Gronlandia provincia

Capud maxi dies ht hor 24

Isla ndia

Norue

Nidrosia

B e

bergen

Nordmannia

Stauager

Capud maxi dies ht hor 20

Capud maxi die ht horas 19

Gothia

Iberuia insula

Suoue regio

Capud maxi die ht hor 18

Capud maxi die ht horas 17

Britani anglicani apostate

Bolgari Slauorum

bula xi

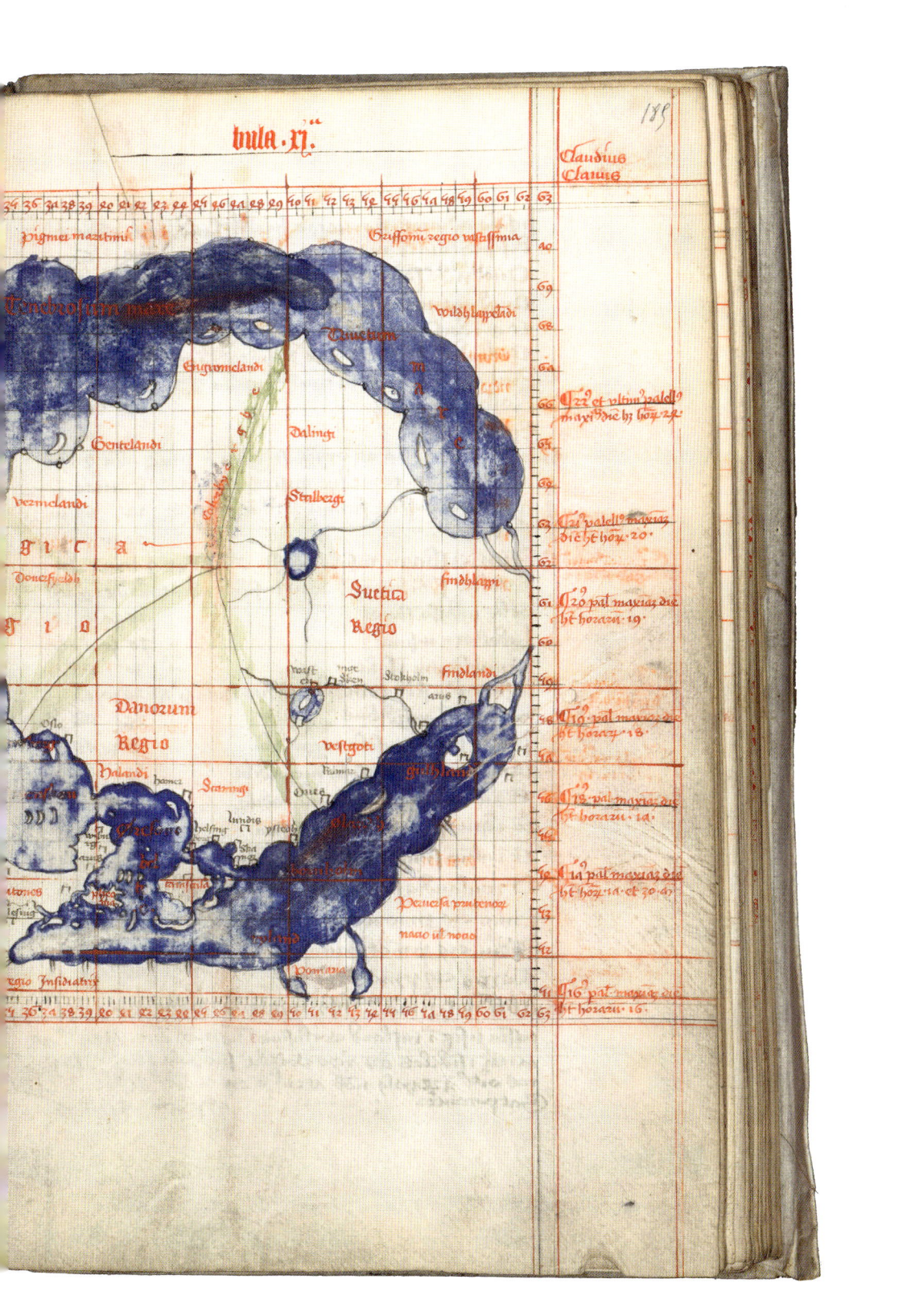

Claudius Clauus

Pigmei maritimi — Griffonū regio vastissima

Tenebrosum mare — wildhlappeladi

Gruwnelandi — Thueum

Gentelandi — Dalingi

Vermelandi — Stalbergi

Donezfyeldz — findlappi

Suetia Regio

Danorum Regio — Stokholm — findlandi

Halandi — Vestgoti — gothlandi

Pomerania — Prauessa pruteno̅r

Scandinavia with Sweden to the east, Denmark to the south and a compressed Norway to the west (FIG. 10). Green lines roughly demarcate the boundaries between the kingdoms – which in fact were in a political union at the time of the map's making – and important pilgrimage sites such as Nidaros in Norway and Vadstena in Sweden provide a religiously informed framework for the map that would have spoken to the priorities of its contemporary viewers.

The Ulm edition of Ptolemy's *Geographia* included a map of Scandinavia by Nicolaus Germanus with information likely originating from a now lost map by Claudius Clavus, though the general outline of this map and Clavus's surviving map kept in Nancy differ significantly from each other.[26] Germanus's map shows a web of islands and peninsular land masses in the North Sea, north of Germany and the British Isles (FIG. 11). The contours of Scandinavia would hardly be recognizable as such if not for the helpful legends on the map, noting, for example,

11 Nicolaus Germanus's map of the Nordic region that was part of the 1482 Ulm edition of Ptolemy's *Geographia*. It perpetuated the confusion surrounding the location and number of places called 'Greenland'.

that the western part of the peninsula is 'Norbegia' and the eastern 'Suecia' and 'Gothia'. More localities are set out on the Danish Island Zealand (where Copenhagen is situated and where Claudius Clavus was educated), though it is a reduced and somewhat distorted Zealand that appears on Nicolaus Germanus's map of Scandinavia. Iceland and the

THE UNKNOWN NORTH   35

peninsula labelled 'Engronelant' and 'Pilappelandt' appear north of the Polar Circle. 'Engronelant' can also be found further south on the northern part of the Scandinavian peninsula. The configuration led to an enduring confusion among Renaissance scholars as to whether there were one or two Greenlands, and whether Greenland was an island or a peninsula. This was an important question for European merchants and seafarers in search of either the Northwest Passage or the Northeast Passage to Asia, since a peninsular Greenland would have needed to be traversed or circumnavigated to reach the Pacific Ocean. For information about the number and location of Greenlands, Ptolemy's geography was little help. Having transformed the way early modern Europeans made maps, his work again became obsolete.[27]

The early printed editions of Ptolemy's *Geographia* were luxury artefacts with meticulously made woodcuts, and many of the maps were beautifully coloured by hand. The edition printed in Ulm in 1482 exhibits vivid blue, red and green colours, and it was printed on expensive paper. In fact, it is possible that the Ulm edition was an excessively lavish production, since Leinhart Holl's printing house filed for bankruptcy after it was published.[28] The result, however, was a collection of maps fit for princes. A copy of the Ulm Ptolemy now in the Bodleian Library was previously owned by Ferdinand and Isabella, the monarchs of Castille and Aragon, who famously financed Christopher Columbus's journey across the Atlantic Ocean in 1492.[29] It is clear, then, that maps like the Ulm Ptolemy were rare and exclusive objects and not accessible to the masses. Still, given the power and influence of sovereigns like Ferdinand and Isabella, luxury maps viewed by a select few had the potential to influence worldviews and decisions that affected many more.

12 The contours of the continental land masses are vaguely recognizable on this map of the climate zones from late-tenth-century Germany. The point of the map, however, was not to present a detailed geographic account, but to illustrate the extension of the temperate, torrid and frigid zones across the Earth.

Map labels (clockwise from top):

REFVSIO OCEANI AB OCCIDENTE IN SEPTENTRIONEM · REFVSIO OCEANI AB ORIENTE IN SEPTENTRIONEM · REFVSIO OCEANI AB ORIENTE IN AVSTRVM · REFVSIO OCEANI AB OCCIDENTE IN AVSTRVM

FRIGIDA SEPTENTRIONALIS
TEMPERATA ITALIA · TANTA · MARE CASPIVM
ORCADES
PERVSTA · RVBRV MARE · MARE INDICVM
OCEANVS
PERVSTA
TEMPERATA ANTIKTORVM
FRIGIDA AVSTRALIS

Quod autem n̅ram habitabilem dixit ab angustam verticibus
        t aduertere
lateribus latiorem in eadem descriptione p̄bare poterimus · Nam quanto longior
      expoul   co
est tropic̄ septemtrionali circulo · tanto zona uerticibus quam lateriby angusti
              t circuli   con
or est · Quia summitas ei ix artum extrem̄ cinguli breuitate trahitur · Deductio
                                                                   em
aūt laterum cum longitudine tropici abutriaq; parte distenditur · Deniq; ueteres
omnem habitabile̅ n̅ram extem̄e clamidi similem ee dixerunt · Itq; omnis tia
in qua ⁊ oceanus ē ad quem uis celeste circulum quasi centron puncti obtinet locum ·
Necessario de oceano adiecit · qui tamen tanto nomine quam sit paruus uides ·
Nam lic⁊ apud nos athlanticum mare lic⁊ magnum uocatur · de celo tamen
despicientibus non potest magnum uideri · cum ad c̄lum terra signum sit
quod diuidi non possit in partes · Ideo aūt terre breuitas tam diligent asserti

ISLANDIA.

*Privilegio Imp. et Belgico decennali
A. Ortel. excud. 1585*

Scala milliarium Islandicorum

ILLVSTRISS. AC POTENTISS.
REGI FREDERICO II DANIAE,
NORVEGIAE, SLAVORVM, GO
THORVMQVE REGI, ETC. PRIN
CIPI SVO CLEMENTISSIMO,
ANDREAS VELLEIVS
DESCRIBEB. ET DEDICABAT.

MAPPING COLD

Conceptions about the northernmost parts of the world were also elaborated in maps that focused on climate. According to ancient climate theory, the world consisted of a set of climatic zones that ran parallel with the equator. Maps that illustrated the climate zones were well known in learned circles in Europe as well as the Middle East from antiquity into the early modern period, and they often accompanied textual manuscripts, such as the fifth-century scholar Macrobius's *Commentary on Cicero's 'Dream of Scipio'*. There were variations in how different scholars interpreted the climate zones, but the general understanding was that the area around the equator corresponded to a torrid zone with such a harsh climate that it was uninhabitable. In contrast, the zones on either side of the torrid zone were believed to be temperate, and these were well suited for human habitation. The zones that encircled the polar regions were, in contrast, believed to be unbearably cold and unfit for people to live in.[30]

As Europeans living farther south learned more about the northern reaches of Europe and Asia, they had to revise the proposition that the northern 'frigid' parts of the world were impossible to inhabit: people were clearly living in the uninhabitable zone. Still, depictions of a cold and inhospitable north continued to have an extra layer of meaning beyond simply showing a harsh climate because climate theory dictated that weather conditions could shape human physiology in detrimental ways. Medical thinkers believed that human bodies and mental faculties were shaped by the climate of the region where a person was born and raised. That made climate not just a question of hot or cold – it was closely bound up with evaluations of human potentiality and with categorizations of us and them, adding a further element to how

13 PREVIOUS SPREAD Made during the coldest part of the Little Ice Age, Ortelius's rendition of Iceland portrays a mountainous island located in a frosty sea. Polar bears occupy the ice floes off the north-eastern coast and large sea creatures are seen in the waters to the south. On the island itself the volcano Hekla is spewing forth lava and ashes.

contemporary audiences would have interpreted maps that portrayed the northernmost parts of the world as a domain of snow and ice.

How, then, have map-makers gone about portraying the climate of the far north on their maps? Nicolaus Germanus and other early European map-makers found it sufficient simply to include a legend describing the sea in the north as *Mare Congelatum*, Latin for 'The Frozen Sea'. Others chose to elaborate on the theme, including depictions of sea ice on their maps. Thus, Olaus Magnus's *Carta Marina* shows a partially frozen Baltic Sea traversed by skiers, and polar bears carousing on ice floes off Iceland's north-eastern coast.[31] A couple of decades later, the famous Flemish map-maker Abraham Ortelius similarly portrayed Iceland as located in a cold climate, partly by including illustrations of ice floes on the map (FIG. 13), and partly by the accompanying textual description which detailed that the letter Q on the map refers to 'Huge and marvailous great heaps of ice brought hither with the tide from the frozen sea, making a great and terrible noise; some pieces of which oft times are fourty cubites bigge.'[32]

Like the many other decorative elements on Olaus Magnus's and Ortelius's maps, the illustrations of ice serve to create a sense of place; in their renditions, the north and Iceland respectively are cold regions with uncontrolled natural forces. Moreover, the depiction of ice on the maps freezes Iceland to the winter season. Olaus Magnus and Ortelius made their maps during the coldest part of the Little Ice Age (*c.* 1300–1850); even then the ice would have melted during the latter part of the summer. Yet, looking at the maps, the ice seems to be a permanent feature of the north. In this way, the medium of the map likely contributed to the conception of a cold north.

However, that the north could be reduced to only a cold place was something that a set of Swedish seventeenth-century scholars and statesmen did not want to concede. Instead, they made new interpretations of climate theory that aligned better with the patriotic sentiments of the early modern Swedish Empire. For example, the text accompanying

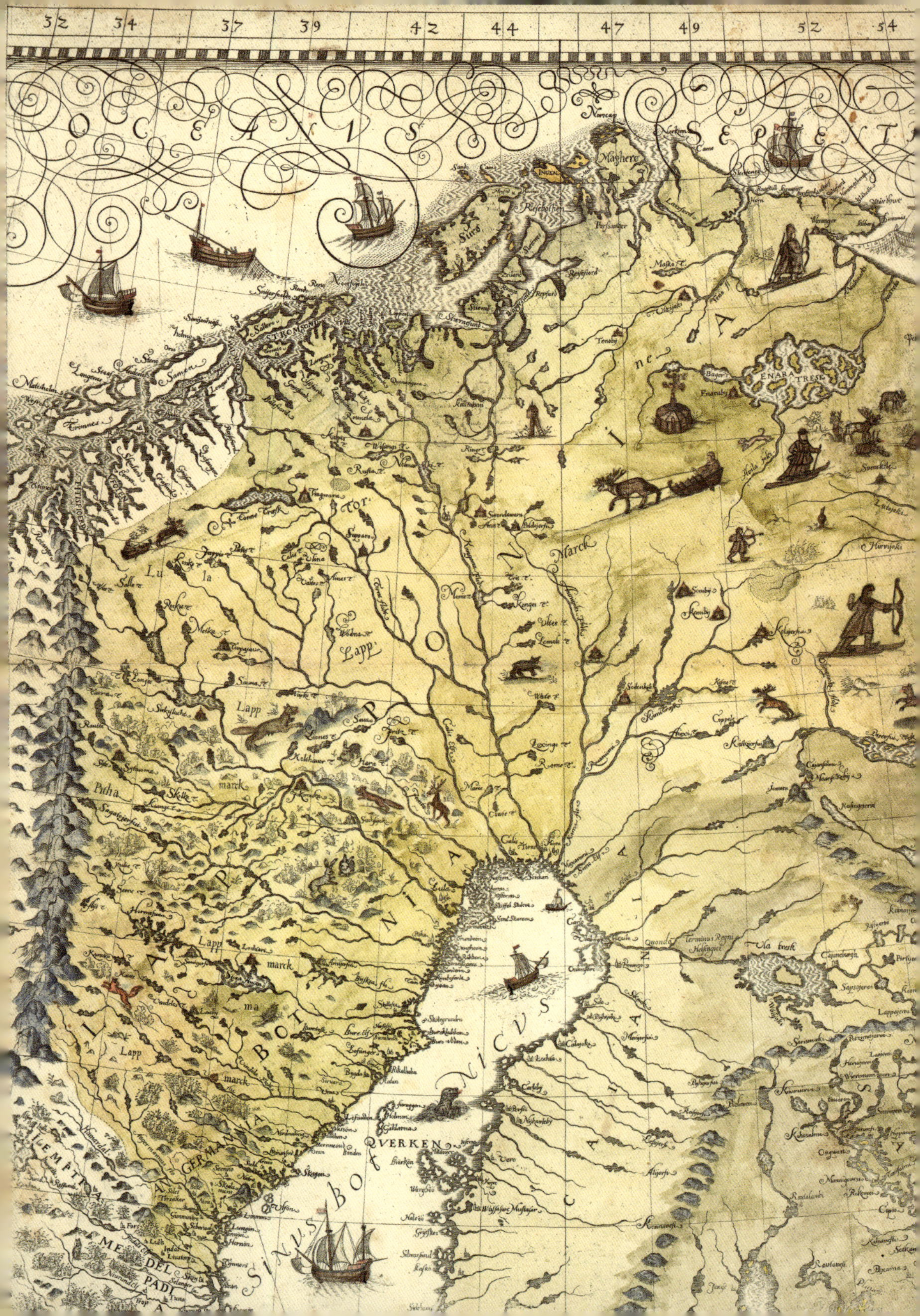

the Swedish map-maker Andreas Bureus's 1626 map of 'the Arctic World' – though the map really only portrays the Nordic countries – notes that Scandinavia's 'southernmost parts up to 60° lat. or less, have a mild and temperate climate; its middle part, between 60° lat. and the Polar Circle, indeed has a climate not so mild in character, but the harshness of the climate is counterbalanced by the fertility of the soil' (see FIG. 2).[33] Bureus had to concede that the area above the Polar Circle had neither a good climate nor good soil, but he was quick to emphasize that an abundance of wildlife compensated for these deficits. Bureus recast the idea of a frigid zone. Now only the northernmost part of Scandinavia was cold and inhospitable; however, it too had value.

In addition, Bureus cunningly emphasized, through the decorative elements of his map, that Sweden was a part of European culture and learning. Instead of showing local customs or stylized images of northern hunters, Bureus chose images of the Roman goddesses Venus, Minerva and Victoria. In the top right corner Neptune is seen riding the waves of the Barents Sea on a chariot drawn by sea monsters.[34] These depictions foreground a counter-narrative emphasizing that the Swedes were not part of the suspect frigid zone, but instead belonged to the temperate – and thereby civilized – world.

This book mostly focuses on map-makers; how the makers of maps have grappled with the difficulty of plotting maps of an inaccessible region. Yet, while a map-maker can have a carefully crafted idea about the region portrayed on a map, a user can easily cut the map in half or give it a new title that changes the original framing. A partial copy of Andreas Bureus's map *Orbis Arctoi* has undergone such a transformation (FIG. 14). A zealous user has coloured the northernmost part of Scandinavia green, and as a result the skiers on the map seem to be

14 This copy of Andreas Bureus's *Orbis Arctoi* has been coloured green. A reindeer is seen pulling a Sami sleigh across the green expanses of the north of Sweden, and other figures are skiing through the grass-like surface of the map.

THE UNKNOWN NORTH   43

The Line of Perpetual Snow, or Region of Congelation
Map of
SWEDEN,
NORWAY, AND LAPLAND,
With numerous interesting & useful details
relating to the Climate Productions &c.
of These Countries by
Lt. Col. HAGELSTAM:
Translated from the Swedish by
ARTHUR DE CAPELL BROOKE, F.R.S.
Explanation of the Terms, Marks &c. occurring in the Map.
The Red Line shows the Author's Route.
SEA
OF
BOTHNIA

skiing through green grass.³⁵ It might be that the colourist wanted to convey a summer view by choosing a green tint for the map, though the motivation was likely also political in nature. In the far north the colouring follows Bureus's original demarcation of the border between Sweden and Norway with a dotted line at Malangen fjord in present-day northern Norway. There was a historical precedent for this demarcation, but at the time Bureus made his map the area was part of Norway. Thus, the border demarcation was a political statement favouring

15 Arthur de Capell Brooke's *Map of Sweden, Norway and Lapland* is a translation and reworking of an even more information-packed Swedish map. In the margins, de Capell Brooke provided information about topics such as the effect of the Gulf Stream on Norway's climate, and which plants grow at what altitudes.

16 Accompanying his travel narrative and map of Scandinavia, de Capell Brooke produced a set of prints to illustrate his journey. The portrayal of a reindeer-drawn sleigh is reminiscent of numerous other depictions of Scandinavian and Sami winter travel, such as portrayed on Bureus's map *Orbis Arctoi*.

THE UNKNOWN NORTH 45

Swedish claims to the north of the Scandinavian Peninsula. With the green tint the colourist has reinforced Bureus's patriotic view, while also adding a new layer to the map, which now seems to belie the idea of a frigid and uninhabitable north.[36]

Over time, and as the ideals of map-making changed, the richly decorated medieval and early modern depictions of the north gave way to maps with, on the whole, fewer icebergs and polar bears. This did not, however, necessarily mean that the maps stopped portraying a cold and harsh northern climate. For example, labels indicating cold areas continued to appear on maps. An interesting elaboration of this is seen in the nineteenth-century British adventurer and travel writer Arthur de Capell Brooke's map of Sweden and Norway.[37] The map accompanied the second volume of de Capell Brooke's travelogue detailing his Nordic sojourn because, according to the author, 'the time occupied in its [the map's] execution, and the difficulties attending it, rendered it impracticable to get it ready in time' for the publication of the first volume.[38] This was, in other words, not a map essential for the would-be Nordic traveller following in de Capell Brooke's footsteps, but more a display of the geography of the region for the armchair traveller. Notations on the map's margins describe the average depth of snow during the winter at various places and give information regarding the altitude at which certain plants can grow (FIG. 15). This is information that was also part of the original Swedish statistical map that de Capell Brooke used for his map, but the translation process has changed the map in subtle ways.[39] For example, de Capell Brooke simplified the terminology of the original map, and the information given emphasizes the cold climate more than does the original map.[40]

In contrast to Olaus Magnus's and Ortelius's maps, de Capell Brooke's rendition of Scandinavia includes information about the changes of weather over the course of the year. Nevertheless, the result is a map of a region characterized by cold. This impression would have been strengthened even more for a reader who also looked at de Capell

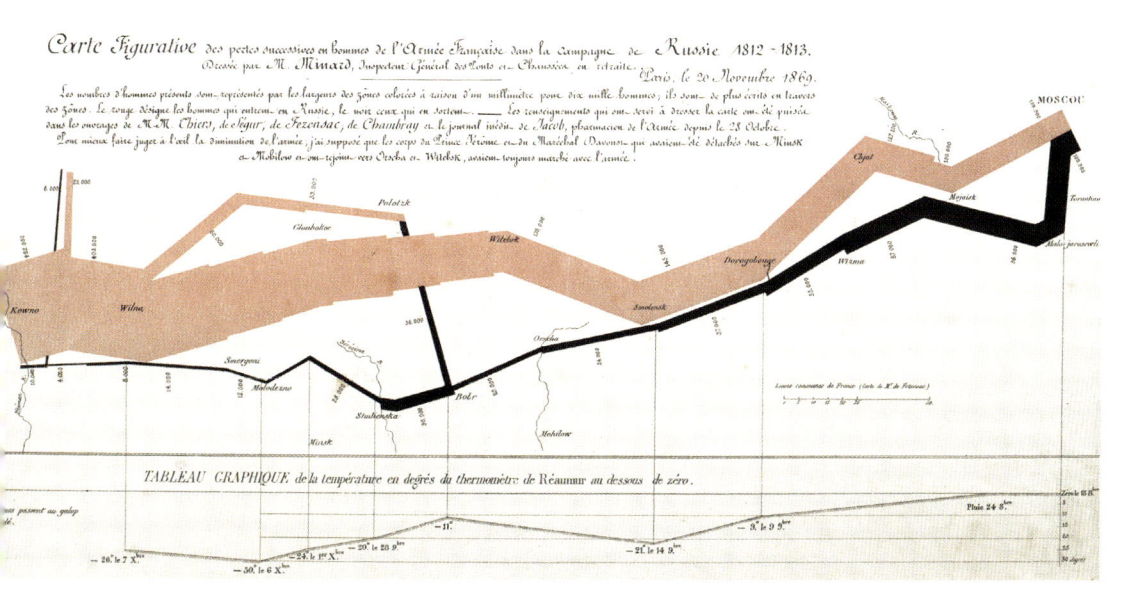

17 Charles-Joseph Minard's map shows the fatality rates that the Napoleonic army suffered during the Russian campaign of 1812–13. The size of the army on the eastward journey is depicted in red, and the dwindling numbers who returned are marked in black, with corresponding temperatures noted at the bottom of the map.

Brooke's accompanying illustrations of his travels through northern Scandinavia, and who read his travel account of what was portrayed as a dangerous journey. Through images and texts, the theme of a cold north occurred again and again in the accounts of explorers travelling through northern latitudes in the nineteenth and early twentieth centuries (FIG. 16). This reflected real-life experiences of the travel writers. However, the portrayal of the natural environment as cold and dangerous could also be used to emulate the stamina and daring mindset of the traveller. To dwell on the difficulties of a journey was a way to garner sympathy and admiration. Simultaneously, the described environment appeared exotic and extreme.[41]

The cold climate of the north has also been used in explicit ways to create a narrative through the map medium. The French civil engineer Charles-Joseph Minard published such a map in 1869, where he made

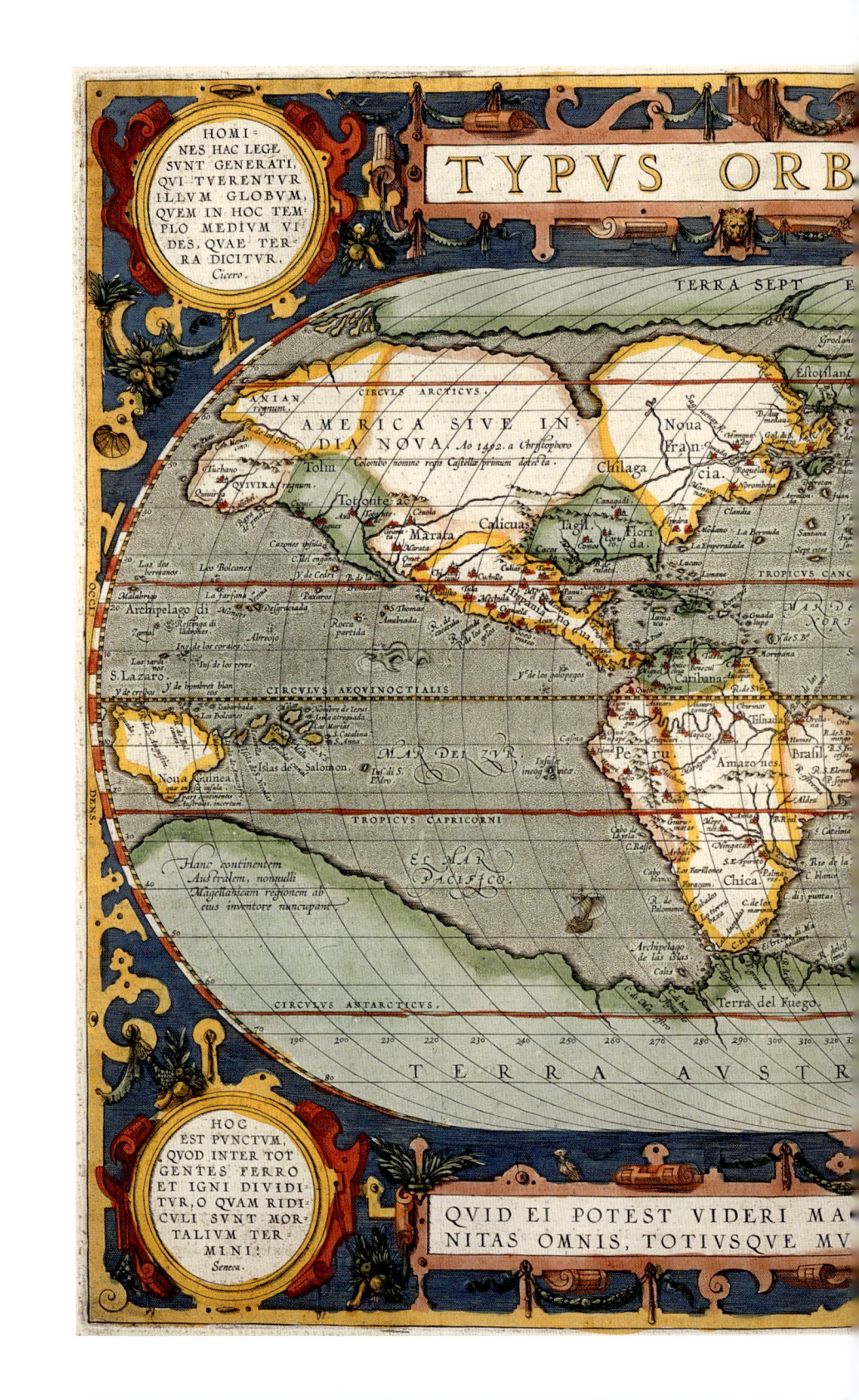

> HOMI-
> NES HAC LEGE
> SVNT GENERATI,
> QVI TVERENTVR
> ILLVM GLOBVM,
> QVEM IN HOC TEM-
> PLO MEDIVM VI-
> DES, QVAE TER-
> RA DICITVR.
> *Cicero.*

TYPVS ORB

TERRA SEPT

CIRCVLVS ARCTICVS

AMERICA SIVE IN-
DIA NOVA.

Ao 1492. a Christophoro Colombo nomine regis Castellæ primum detecta.

TROPICVS CANC

CIRCVLVS AEQVINOCTIALIS

MAR DEL ZVR

TROPICVS CAPRICORNI

EL MAR PACIFICO

Hanc continentem Australem, nonnulli Magellanicam regionem ab eius inventore nuncupant.

CIRCVLVS ANTARCTICVS

Terra del Fuego

TERRA AVSTR

> HOC
> EST PVNCTVM,
> QVOD INTER TOT
> GENTES FERRO
> ET IGNI DIVIDI-
> TVR, O QVAM RIDI-
> CVLI SVNT MOR-
> TALIVM TER-
> MINI?
> *Seneca.*

QVID EI POTEST VIDERI MA
NITAS OMNIS, TOTIVSQVE MV

TERRARVM.

EQVVS VEHENDI CAVSA, ARANDI BOS, VENANDI ET CVSTODIENDI CANIS, HOMO AVTEM ORTVS AD MVNDVM CONTEMPLANDVM.
Cicero.

...LIS INCOGNITA.

...S NONDVM COGNITA

...N REBVS HVMANIS, CVI AETER= ...TA SIT MAGNITVDO. CICERO:

VTINAM QVEMADMODVM VNIVERSA MVNDI FACIES IN CONSPECTVM VENIT, ITA PHILOSOPHIA TOTA NOBIS POSSET OCCVRRERE.
Seneca.

a direct connection between climate and danger.[42] The map shows the hardship of Napoleon's army during its Russian campaign of 1812–13 and the route that it took from the Polish–Russian border eastward, and then back again (FIG. 17). The size of the army is visually connected on the map to, first, the continually increasing number of casualties, and, second, to the low temperatures experienced by the army on their return journey. It is a stark depiction of the infamous military disaster, and an innovative way in which a geographical map can be used both to illustrate a sequence of events and to describe the climate (and by extension the character) of a region. Published in the same year as Leo Tolstoy's novel *War and Peace*, which also takes Napoleon's Russian campaign of 1812 as its topic (albeit taking a Russian viewpoint on the conflict), Minard's map should be seen as part of a memory culture that developed in nineteenth-century western Europe and Russia, creating a narrative of the hardships of the war.[43]

A HISTORY OF THE UNKNOWN NORTH

Abraham Ortelius's map of the world became one of the most influential world maps of the late sixteenth and early seventeenth centuries.[44] The map was published at a time when geographical information was rapidly expanding, and when the European appetite for new geographical knowledge – to trade, to dream and to subjugate – seemed insatiable. This posed a problem for Ortelius and his fellow map-makers. How were they to portray the areas of the world that were not yet known to them, while still maintaining that they provided the best and most accurate information available?[45]

One response to this issue was to divide geography into that which was known in the present and that which was as yet unknown. Ortelius

18 PREVIOUS SPREAD Ortelius's famous world map, published in 1570, is an account of the newest geographical information available, as well as an admission that parts of the world were as yet unknown.

labelled the northernmost reaches of the Earth the 'northern unknown land' and the southern (rather oversized) land mass 'the southern land [which is] not yet known' (FIG. 18). It might seem strange that Ortelius bothered to draw the contours of land which he also claimed were unknown. However, while the existence of these land masses was not confirmed, there were still conjectures about what might lie beyond the part of the world where Europeans had sailed. By including the unknown northern and southern continents, we might say that Ortelius hedged his bets, indicating to his audience that he was aware that his view of the world would one day become outdated.

As the number of expeditions north increased, a common strategy among map-makers was to include information on their maps indicating when different areas had first been reached by explorers. In this way they could highlight that they had included up-to-date information. Thus Ortelius noted across North America that this land was 'discovered by Christopher Columbus in 1492'.[46] By emphasizing that land was 'unknown' and later 'discovered', Ortelius and other early modern map-makers emphasized a Eurocentric perspective on world geography: geographical information had a history, and that was the history of European discoveries.

In addition, including notes about geographical 'discoveries' became a way for explorers to publicize the new information gained from polar expeditions, and for states to claim newly explored lands. An example of this can be seen in a map published in 1758 by the Academy of Science in St Petersburg, which shows the northernmost parts of Russia and America (FIG. 19).[47] The map was compiled by the German–Russian scholar, explorer and map-maker Gerhard Friedrich Müller; it presents information gained by Russian expeditions east during the

19 OVERLEAF This map, compiled by Gerhard Friedrich Müller, was made to showcase Russian claims in the northern Pacific Ocean while also providing new geographical information. Along the north-western coast of America, legends place Russian discoveries in a context of earlier European expeditions.

NO...
DES DEC...
VAISSEAUX R...
DE L'AMERIQ...

Dressée sur des...
Qui ont assisté...
Connoissances, d...

A St. Peters...

MER GLACIALE

MER D'OCHOZK
appellée LAMA
par les Toungouses

MER DE
KAMTSCHATKA

ISLE DU NIPHON

ISLES DES KURILES

MER

AMERIQUE SEPTENTRIONALE

BAFFIN'S BAY

DETROIT de BAFFIN

DETROIT de HUDSON

HUDSON BAY

LABRADOR

NOUV. SUD GALLES

NOUV. ALBION

PARTIE DE CALIFORNIE

SUD

preceding decades, which Müller himself had taken part in. On the face of the map, Müller plotted the route that the Danish explorer Vitus Bering had travelled along the Kamchatka Peninsula in 1728, an assignment directed by Tsar Peter the Great. Müller also sketched out the contours of the Bering Strait, named after Vitus Bering, and noted the itineraries of Russian explorers along the north-western coast of America in relation to earlier western European expeditions. Thus, Müller pointed to a stretch of coast encountered by the Russian navigator Aleksei Chirikov in 1741, next to references to a sixteenth-century Spanish expedition led by Juan de Fuca. A conjectured system of rivers connects Hudson Bay to the Pacific Ocean, postulating the possibility of a passage between Asia and Europe.

As a result, Müller's map provided new information about the geographical outlines of the north-east of Asia and the north-west of America to a European market, while also emphasizing Russia's claims to these areas. Behind these claims lay an effort to document the north-east of Russia's hinterland, a project which in turn included expeditions sent out by the authorities, local guides, and explorers and map-makers from several different countries.[48]

Map-makers and map publishers continued to mark discoveries on maps of the north into the twentieth century. The practice is seen on the maps of the North Pole from the popular German atlas *Stielers Hand-Atlas*, where lines and legends show the successive discoveries of explorers traversing the north (FIGS 20 & 21). *Stielers Hand-Atlas* went through numerous continually expanding and widely distributed editions during the nineteenth and early twentieth centuries, becoming a staple atlas in German schools and institutions. The publishing house Justus Perthes in Gotha, which produced the atlas, prided itself on accuracy and adherence to new scientific standards.[49] As such, the map of the northern polar region was continually changed to include the most up-to-date information. The North Pole map appearing in the ninth edition of the atlas, its maps having been prepared in the period 1900–1905, noted a number of recent explorations in the Polar Sea:[50] for example, that the Norwegian explorer Fridtjof Nansen reached 86°14′N in April 1895, that his Italian counterpart Umberto Cagni reached 86°34′N five years later, and that the American explorer Robert E. Peary reached the northernmost tip of Greenland in 1902. While none of these expeditions reached the ultimate goal of 90°N, all the notations set records that were followed by a public fascinated by the idea of explorers taking on the challenge of the farthest north. Thus, the notations representing recent expeditions north both filled the role of creating a narrative of the expansion of geographical knowledge and became a spatially presented hall of fame for polar explorers.

When a new 10th edition of *Stielers Hand-Atlas* appeared in the 1920s, the race for the North Pole was seemingly over, with the map declaring that Peary had reached the North Pole on 6 April 1909.[51] However, what the map does not show is that Peary's claim was hotly debated at the time of the publication of the atlas (and doubts as to whether he actually got to the North Pole remain today). By selective inclusion and exclusion, the map portrays information as less contested than it was.

Parallel to accounts of new records and discoveries, the public interest in historical atlases grew in Europe.[52] Historical atlases provided viewpoints of historical periods and events, and some of them also paid particular attention to the status of geographical knowledge at different points in time. Perhaps one of the most memorable examples of this

20 OVERLEAF This map for the ninth edition of *Stielers Hand-Atlas* was prepared in 1900–1905. It provides an overview of the North Polar Sea and surrounding land. The margins of the map are occupied by maps of Arctic islands, the ocean currents (*lower left*) and the distribution of the Northern Lights (*lower right*).

21 FOLLOWING SPREAD The North Pole map from the tenth edition of *Stielers Hand-Atlas* states that Robert E. Peary reached the North Pole in 1909, though the claim was contested already at the time of the map's making.

THE UNKNOWN NORTH

NORD-
Entw
BERICHTIGT VON H. HABENICHT.
NÖRDLICHE DURCHFAHRT zwischen ELLESMERE Ld. und GRÖNLAND
Mafsstab 1:10 000 000
PENDULUM-INSELN
Sabine I.
Ostküste von GRÖNLAND
BERING-STRASSE
TSCHUKTSCHEN
SEESTRÖMUNGEN
Nord Pol
ALASKA
BRITISCH
NORD
AMERIKA
Beaufort See
Banks Ld.
Victoria Ld.
Melville I.
Parry In.
Ellesmere Ld.
Grinnell Ld.
Hudson
Nord Southampton I.
Melville H.I.
Cockburn Ld.
BAFFIN LAND
LABRADOR
DAVIS STRASSE
GRÖNLAND
König Wilhelm Ld.
Hudson Ld.
Kg. Christian IX Land
ISLAND
Ost-Grönland See
Erklärung
Forschungsreisen
Gezeichnet v. C. Bohmer 1903.
Breitenmafsstab 1:20 000 000
GOT

KARTE.

haus.

Breitenmafsstab 1:20000000

Englische geographische od. nautische Meilen 60 = 1 Grad

Leitung: Dr. H. Haack

NORDKAP
Maßstab 1:2 000 000
NOWAJA SEMLJA
Maßstab 1:10 000 000
GODTHAAB u. UMGEB.
BÄREN-I.
JAN MAYEN
SPITZBERGEN
Maßstab 1:7 500 000
Nordostland
Spitzbergen
Neu-Sibirische I.
NORDENSKIÖLD-SEE
JAKUTSK
SIBIRIEN
JENISSEISK
WEST SIBIRIEN
KARA-SEE
BARENTS-SEE
Lappland
Kola
Omekon
Stich Kögel Reichenbecher

kind of historical atlas is Edward Quin's *An Historical Atlas, Containing Maps of the World at Twenty-one Different Periods* (1830). Quin was an English barrister with a particular interest in history and geography. He objected to the lack of pedagogy in the teaching of history, where students needed to consult many different books to gain an understanding 'of the bold and leading outlines of History'.[53] Instead, Quin published a historical atlas with maps showing the world as it was known to different societies at particular moments in time. In order to emphasize that the rest of the world existed, albeit unknown to a society, Quin plotted all of his twenty-one maps in the same size and scale while covering those areas that were unknown at a particular time in thick billowing black clouds.

Quin's first map shows the known world in the year 2348 BCE, which in a religiously informed view of history was understood to have been the year of the biblical Flood (FIG. 22). The account in Genesis describes how only Noah and his family survived the floodwaters; reflecting this, Quin's map is almost entirely black, with only a small peephole showing the region around Mount Ararat in present-day Turkey, where Noah, his family and the assorted animals were thought to have been stranded after the floodwaters receded. On this map, the known north is located a few miles north of Mount Ararat.

In contrast, Quin's last map is entirely devoid of clouds (FIG. 23). It portrays the world in Quin's own time – that is, in 1828. Emphasizing the progress of knowledge – and, implicitly, the progress of civilization – Quin argued that the entirety of the globe was now known. Still, he conceded that there were areas that, although known, were not yet fully surveyed. Moreover, he found some parts of the Earth difficult to classify:

22 As an aid for the teaching of history and geography, the English barrister Edward Quin constructed a set of maps that show the state of geographical knowledge at different points in time. His first map illustrates how only the environs of Mount Ararat were known at the time of the Deluge.

B.C. 2348. THE DELUGE.

A.D 1828. END OF THE GENERAL PE

23 The last map in Quin's atlas shows the state of geographical knowledge in his own time. No more black clouds shroud parts of the world from view, yet the northern outlines of America remain decidedly vague, indicating that these regions were not yet fully surveyed.

> The colours we have used being generally meant to point out and distinguish one state or empire from another, and to shew their respective limits and extent of dominion, were obviously inapplicable to deserts peopled by tribes having no settled form of government, or political existence, or known territorial limits.[54]

Instead of colouring these difficult areas individually, Quin gave them all the same 'flat olive shading'. With regard to the northernmost parts of the globe, the result is an almost uniform colouring above 70°N. Not only does this create a much-simplified view of the northernmost regions of the world in the early nineteenth century; Quin has also used the same colouring for all areas and time periods that he had difficulty classifying. This means that he has confined his account to, on the one hand, the areas of the Earth with a history, and, on the other hand, the olive-coloured ahistorical world of what he called 'barbarous and uncivilized countries'.

Quin's historical atlas is a vivid example of how the location and the extent of 'the north' have changed over time, and it is illustrative of the idea that geographical knowledge has a history. More importantly, though, Quin's presentation of the growth of geographical knowledge and its connection to politics, even to conceptions of 'civilization', is a stark reminder that maps both reflect and reinforce power relations between different societies. When used as didactic tools – as Quin's maps were intended – they play a part in shaping our knowledge of the world.

MAPPING THE NORTH FROM AFAR

Viewing the north from afar, the map-makers we have met in this chapter suggested that they knew something about the northernmost reaches of the Earth, even if what they knew was little more than an acknowledgement of a lack of knowledge or that it was a cold place. In the late medieval and early modern period, this was not a feature peculiar to maps of the north. After all, the Renaissance versions of

Ptolemy's world map are equally silent about the southernmost parts of the world as they are of the northern parts. However, whereas European map-makers over time came to consider that they had gained reliable information about most other regions of the world, parts of the northernmost reaches of the world continued to keep the label 'unknown' into the eighteenth, nineteenth and even twentieth centuries.[55] The northern known unknown was there on maps, and this meant that map-makers had to relate to it. Whether this meant leaving blanks, projecting politics and religion north, or portraying a cold and barren wilderness, the resulting maps played a part in articulating what the north was.

statū rapidissimo cursu uenīt duceret illos ad locū eius
dem īsule usq̄ dum nauis resedit. nō longe a t̄ra. erat
nam̄q̄ ripa illi9 tante altitudinis īta ut suum ui
tcem illi9 uix potuissent uidere ⁊ erat color illi9 sicut
carbonū ⁊ mīre r̄etundīs sicut nur. Annus quī
qui remāsit cō tr̄ib9 fr̄ib9 q̄ fisecuti sūt sc̄m .b. de
suo monastio exsiliuit foras de naui ⁊ cepit dea
bulare usq̄ ad fudamētū ripe ⁊ cepit clamare
dicens. Ve m̄ pat̄ pdor̄ a uob̄ ⁊ no h̄eo pt̄acē ut
possim uenire ad uos. fr̄es cōfesti nauī a t̄ra
reducebant ⁊ clamabat ad d̄m dīces. miserēe
nob̄ dr̄e miserere nobis. S̄c̄ nō ueniālui9 pat̄ cū
sui9 soci9 aspiciebat quoīn ducebat ille ī felu
a m̄ltitudine demoniū ad torm̄ta. ⁊ quoīn incende
bat9 m̄o illos. atq̄ dicebat. Ve tibi fili q̄ ī uīta cu
a m̄cuita tale fīne. Iterū arripuit illos p̄sp̄ uē
tus ad australē plagam. Cū aspexissent a longe
retro illā īsulā uiderūt montē discoopirī a fumo
⁊ a se sum̄mēte flāmas usq̄ ad ethera. ⁊ iterū ad
se easdē flāmas respirate. Ita ut tot9 mōs usq̄
quasi un9 rog9 apparuisset. Cūerat sc̄s .b. cū nauigas
set contra meridiem ut septem dies annuit illis
īmare quedā formula. quasi ho̅is sedētis sup
petrā ⁊ uelū ant̄ illū a longe q̄ mēsura uni9
sagi pendēs int̄ duas forcellas fronte ⁊ sic a
gitabat flu̅ctib9 sicut nauīcla solet cū iudicat
a turbine. alii ex fr̄ib9 dicebāt q̄ auis ēt. alii

MAPS & FICTIONAL TRAVEL

Before the age of modern surveying techniques, the accounts of travellers constituted one of the most important – if not *the* most important – sources for map-makers when making maps of the northernmost reaches of the world. Readers have long been drawn to accounts of adventures, hardships and marvels experienced by seasoned travellers in faraway lands. A good travel journal gives the reader a chance to see the world anew, without ever having to leave home. However, while present-day readers can most likely verify an account or find a critical review after a search online or a trip to the library, map-makers in the past had to contend with a greater degree of uncertainty regarding the veracity of the accounts they used to make maps. When travel itself was more difficult and lines of communication slower, it was even more difficult to verify information. This meant that when making a map of a region about which they lacked first-hand knowledge, southern map-makers had to decide which travel accounts were credible and how to translate the often vague information they provided onto the map.

Many travel accounts have been verified in later time periods; others, it has now been concluded, were largely invented. Often, a certain element of ambiguity remains in many of these accounts: the journeys might have taken place, or they might have been figments of human imagination. The veracity of an account could be of vital importance to readers in some situations, yet stories could also be read more as allegories, never intended to reflect real events and places.[1] Regardless, in many cases they made their way onto maps, and stayed there.

A close look at these travel accounts of mythical journeys and the ways in which they were, or were not, considered to be good sources of information for mapping can reveal much about how knowledge of the north was formulated and understood to be credible at different moments. The previous chapter told the story of how map-makers over time came to know more about the broad geographical outlines of the north, and how they filled those outlines with ideas about what the north was like. This chapter turns that story on its head and instead focuses on four specific journeys that never took place, and how information from them made it onto and off maps of the north. The accounts of these journeys may have been intended to appear to their contemporary readers as true descriptions of actual events, as moral parables or as entertaining fables. Common to all, though, was that they became important for the mapping of the north. They show how information from fictitious travel accounts made its way onto the maps

24 & 25 PREVIOUS SPREAD & BELOW The margins of this fifteenth-century manuscript are filled with drawings that illustrate the adventures of the Irish monk Brendan of Clonfert in the North Sea, including an image of an erupting volcano and another of Brendan mistaking a whale for an island.

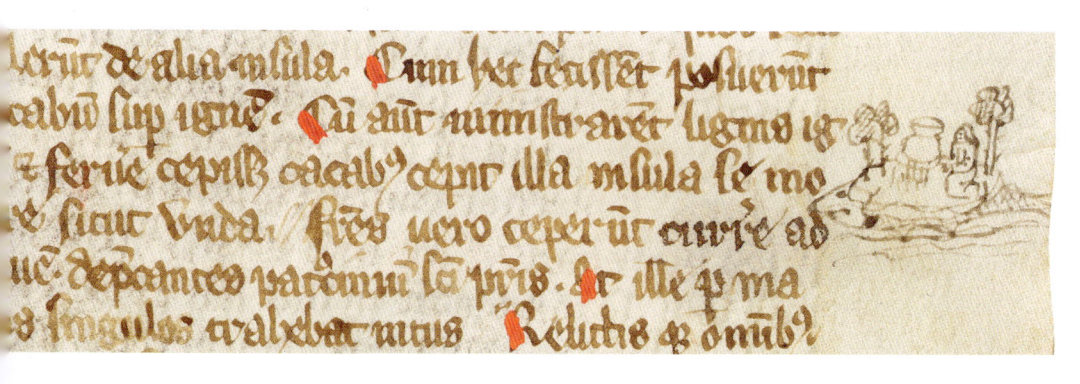

of the great publishing houses of Europe and became integral parts of the cartographic imagination of the north.

SAINT BRENDAN AND THE WHALE THAT WAS AN ISLAND

As the story goes, the monk Brendan of Clonfert set sail from his native Ireland sometime in the first half of the sixth century. His aim was to find the 'Land of Promise of the Saints'; that is, the place where it was believed that the earthly paradise was located. It will come as no surprise that it was a long and difficult journey that awaited Brendan. He and his companions sailed in a small ship across rough seas for seven long years. Each place they arrived at was stranger than the one before it. They landed on an island where drinking from a well put them to sleep, on another island where they feasted on giant sheep, and on a third where the birds were fallen angels in disguise.[2]

Around Easter, Brendan's party came to a small island where the companions, with some relief, were able to debark the ship and celebrate Easter Mass. After they had prayed, they turned to more mundane matters and prepared a fire to cook a meal. As soon as the fire was lit, they started to feel the entire island shake. Before long, the land under their feet began to sink. They had anchored not on an island, but on the back of a resting whale.

Brendan, who was a monk and churchman from Ireland, later became recognized as a saint. Apart from his purported travels on the North Sea, he is known for having founded monasteries, not least the monastery and cathedral in Clonfert, County Galway. The account of Saint Brendan's journey, entitled *Navigatio Sancti Brendani Abbatis*, was most likely composed in the eighth or ninth century.[3] Given the number of surviving copies of the account, clearly generations of readers found it fascinating and instructive. Some of the early manuscripts included illustrations of Saint Brendan's hardships during his years of travel. Among these, an early-fifteenth-century manuscript at the Bodleian Library (FIGS 24 & 25) includes what is believed to be the earliest preserved illustration of an erupting volcano, and, of course, the suggestive image of the campfire on the back of a whale.[4]

In its episodic composition and narrative drama, the account of Saint Brendan's travels follows both early Irish travel accounts, or *immram*, and a tradition of medieval accounts of monastic piety expressed in allegory. The monk's progress through the physical world mirrors a spiritual journey towards salvation. When Saint Brendan sets out on his journey, he must prove that he is worthy in order to reach the Land of Promise of the Saints. The structure of the tale, with recurring feast days and periods of fasting, would also have reminded the medieval audience of the patterns of monastic life. As a reference to what was at stake in the quest for paradise, we are told that three of Saint Brendan's companions died after having committed sins.

The main purpose of the account of Saint Brendan's travels was thus not primarily to describe the various islands, natural phenomena and sea creatures in the North Sea, but to guide the Christian towards a pious life and to instruct on the foundations of monasticism. The value and purpose of such an account did not rest primarily on whether the fantastic islands described were actual places. Instead, the main concern was the moral lessons one could draw from the episodes. Nevertheless, both medieval and later audiences have found the account of Saint Brendan of Clonfert's travels to be realistic – at least in part – and fascinating in terms of propositions about actual geography, and so there is an ongoing debate on whether the settings in the account can be found in the real world. For instance, the island with an abundance of sheep has been claimed to be the Faroe Islands.

In line with this, many medieval maps came to include various elements taken from Saint Brendan's account. Most persistently, islands

26 The Hereford *mappa mundi*, made around the year 1300, includes one of the earliest references to 'St Brendan's islands' on a map. Rather than placing Saint Brendan in the North Sea or northern part of the Atlantic, the map-maker locates 'St Brendan's islands' further south (*bottom right of the map*), identifying them as the Canary Islands.

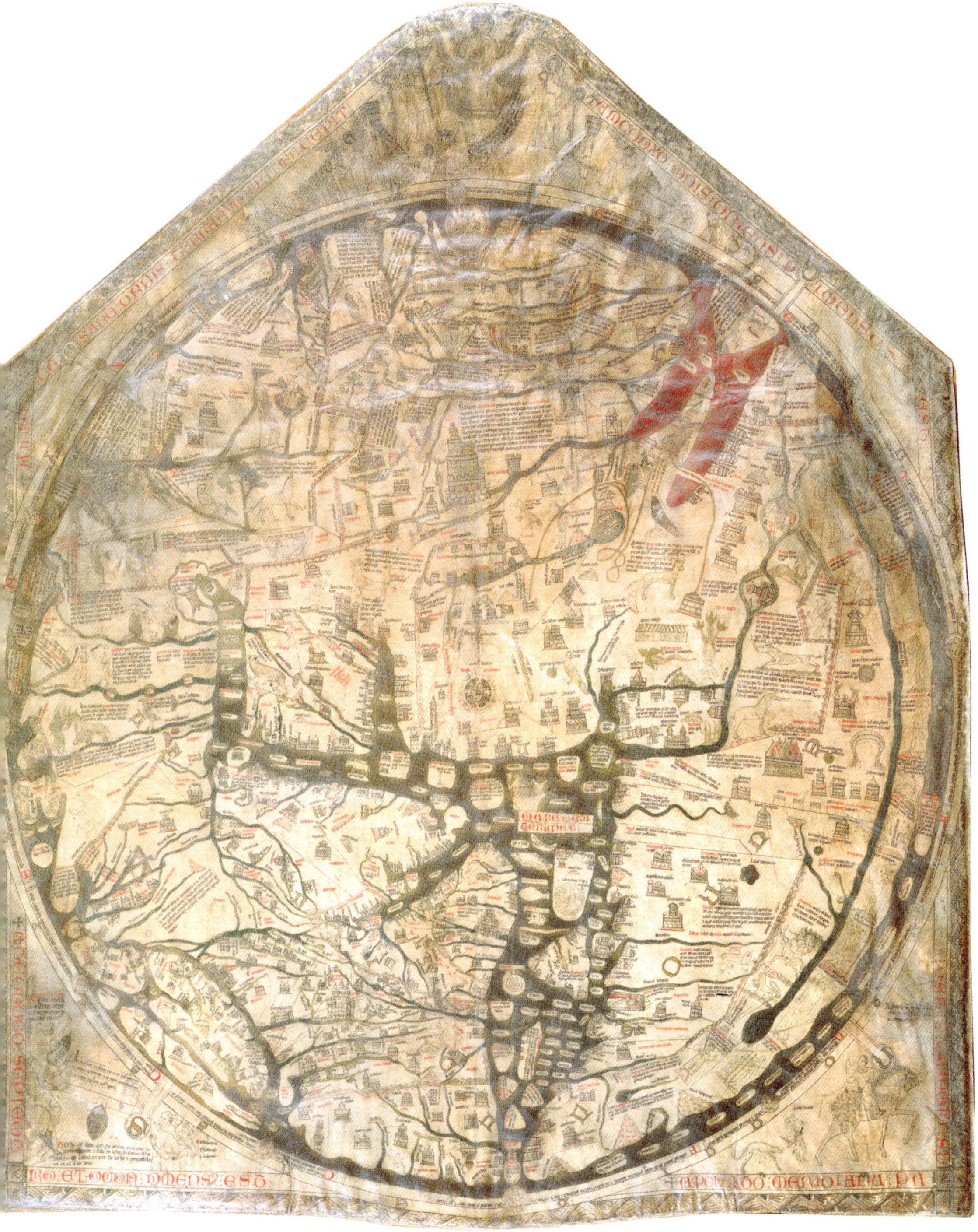

AMERI-
CAE SI-
VE NOVI
ORBIS
PARS

SEP

Groelandt

Neun prom:

Groen: landt.

Estotilant

OCEANVS BO

OCCIDENS

Islant olim Thule

OCEANVS DEVCALI

Drogeo. Dus Cirnes Gallis

OCEANVS OC-
CIDENTA-
LIS

Frisland

Podahila

Hirta

Scetland insi

Hebudes insule

Orcades

SCOTI

S. Brandain

HIBER
NIA

ANGLI

Brasil

y Verde.   Demar

Scala mill: Germanicorū
10 20 30 40 50 60 70 80 90 100

OCEANVS BRITANNI

SEP
TEN
TRI
ONA
LIVM RE
GIONVM
DESCRIP.

named after the saint began to appear on maps. However, as the account, quite understandably, was somewhat vague as to the actual location of the islands that the monk had visited, places as varied as the Canary Islands, Maderia and islands in the North Sea were given the name 'Saint Brendan's Island'. The Hereford *mappa mundi*, one of the largest preserved medieval world maps, thus included a set of St Brendan's islands in the world-encircling ocean, but located them further south off the north African coast.[5] This variety of St Brendan's islands on different maps would not necessarily have troubled medieval audiences. Neither was it unusual for actual locations, events and places that were religiously significant to be placed side by side on a map. As we saw in the previous chapter, the central purpose of the medieval *mappae mundi* was to symbolically represent the world, and this did not necessarily include a sharp distinction between myth and actual locations.

World and regional maps from the sixteenth century onwards began to shed many of the islands and other semi-mythical elements that had occupied the fringes of the medieval world maps. This was partly due to changing ideas about what religious and cultural elements to put on a map, and partly due to increases in travel, which led to a better understanding of where islands actually were located. Still, islands named after Saint Brendan lived on a while longer. For example, in 1570 Abraham Ortelius included a small island called 'St Brandain' on his map of the northern parts of Europe.[6] Over a century later the French cartographer Guillaume Delisle still felt that it was at least worth mentioning the 'fabled' Saint Brendan's Island – this time located west of the Canary Islands – on his map of West Africa.[7] As time went on, references to Saint Brendan disappeared from the face of the map. He

27 PREVIOUS SPREAD The geographical contours of northern Europe are clearly recognizable on Ortelius's map from 1570. Still, the map includes features that later periods concluded were more myth than fact, such as the islands of 'Frisland', 'Brasil' and 'St Brandain'.

28 Olaus Magnus comments that 'Seamen who anchor on the backs of the monsters in belief that they are islands often expose themselves to mortal danger.' It is hard to disagree with that statement. Detail from *Carta Marina*, 1539.

has nonetheless lived on in the popular imagination to the present day as a symbol of the traveller in the north; for example, he is the patron saint of travellers and sailors. It is perhaps more surprising that he is also at times named the patron saint of whales.[8]

The suggestive image of Saint Brendan's meal on the back of a whale became a popular illustration on early modern maps. For instance, when Olaus Magnus made his map of the Nordic world, he chose, among other illustrations of maritime adventures, the episode of Saint Brendan making a landing on a whale. Depicted just south of Iceland on *Carta Marina* (FIG. 28), a ship is made fast to the side of a whale and two men are seen precariously balancing a cauldron over a

MAPS & FICTIONAL TRAVEL

29 In 1621, the Benedictine monk Caspar Plautius published a collection of travel accounts to bolster Catholic exploration of the New World. He included a print showing Saint Brendan's ship stranded on the back of a whale, and Saint Brendan himself holding Mass while the sea creature looked on.

Is: S. Brandano.

Fortunatæ

Cabo de No:

Af

fire on its back.[9] This depiction of cooking on the back of a whale references the tradition of Saint Brendan as a literary theme, and it also connects to a longer tradition of purported encounters with large sea creatures, going back to the biblical story of Jonah and the whale, and into the modern era and J.R.R. Tolkien's poem about the sea creature Fastitocalon, described as 'An island good to land upon'.[10] A parallel theme is the legend of the menacing and many-armed Kraken, a sea creature that would drag ships into the depths of the North Sea. In both cases, the idea of a large and mysterious sea-dwelling monster has taken hold of literary (and, in recent years, cinematic) imagination.[11] Apart from a certain tantalizing exoticism, it is likely that the tales of large sea creatures became popular because they reflected the concerns and experiences of fishermen and sailors in northern waters. Thus, reports of Kraken sighting came from Norwegian sailors who, according to one eighteenth-century commentator, 'spoke with one voice and no contradiction' about the dangerous potential of the monster that lurked in the waters they had to sail.[12]

One of the most striking illustrations of the idea that Saint Brendan mistook the back of a whale for an island appears in the Austrian abbot Caspar Plautius's *Nova typis transacta navigatio novi orbis Indiae occidentalis*, published in 1621.[13] In a beautiful illustration, Saint Brendan is seen holding Easter Mass on the back of the whale, while his ship is stranded on the tail of the animal (FIG. 29). Caspar Plautius, himself a Benedictine monk, took the medieval account of Saint Brendan's adventures and published it together with other illustrations and travel accounts describing journeys to America undertaken by Catholics. In so doing, Plautius repurposed the medieval legend and claimed it as one in a long series of explorations in the North Sea conducted by Christians loyal to the papal seat in Rome. Like Olaus Magnus's *Carta Marina*, Plautius's publication was part of the religious and political positioning of the Counter-Reformation.

In these later maps and illustrations, the instructions about monastic life that were integral to the original account of Saint Brendan's journeys were lost. Instead, the account had become a source for map-makers trying to find information about the location of islands in the Atlantic Ocean. The geographical information in the original travel account was vague, as attested by the varying interpretations of where 'Saint Brendan's island' was located. Map-makers also used the account, through depictions such as the episode of mooring on the back of a whale, as a symbol of the wondrous nature of the north.

TWO VENETIAN BROTHERS SAILING TO A NEW WORLD

It is not only the island of Saint Brendan that catches the eye of the modern observer of Abraham Ortelius's map of the north of Europe. The northern part of the Atlantic is, in fact, littered with strange islands. Just south of Iceland – called Islant or Thule by Ortelius – appears a compact island labelled 'Frisland'. This island has several named ports and cities, and it is surrounded by a band of smaller islands. West of Frisland we find the island Droego and to the north the small island Icaria, as well as a part of America named Estotilant. Ask anyone who has travelled in this region, or looked at satellite images of it: there are no islands corresponding to Frisland, Droego or Icaria in the North Atlantic. Where, then, did these isles come from?

In a volcanic region such as that surrounding Iceland, where the tectonic plates of Europe and America meet, it is not implausible to think that new islands can appear or disappear. After all, the island Surtsey appeared south of Iceland as the result of a volcanic eruption as recently as 1963. From Saint Brendan's account, we also learned that whales can appear as islands. However, it is not volcanoes or whales that are responsible for the islands of Frisland and Droego. Instead, the source is a Venetian travel account.

When Abraham Ortelius placed all these islands on his map, he was using information from an account and map published in Venice a few

SEPTENTRIONALIVM PARTIV

years earlier, in 1558. That publication was written by the Venetian statesman Nicolò Zen the Younger and it described the purported adventures and explorations of Nicolò Zen's ancestors, two Venetian brothers Nicolò (the Elder) and Antonio Zen, who were said to have travelled the North Sea in the last decades of the fourteenth century.[14]

Nicolò Zen's account describes how his predecessor and namesake Nicolò Zen the Elder had lost his way on a sea voyage to the British Isles in the year 1380, and how he ended up on an island far north, called Frisland, as the guest of Prince Zichmni, who governed the dutchy Sorant or Sorano on this island (marked on the southernmost tip of Ortelius's rendition of Frisland). The account details how Nicolò Zen the Elder soon made himself at home and wrote back to

30 Nicolò Zen's 1558 map of the northern parts of the Atlantic Ocean drew on contemporary knowledge of the region, for example including the outline of a large northern peninsula named Engronelant. However, the map-maker also added the fictitious islands Frisland, Droego and Estotiland.

MAPS & FICTIONAL TRAVEL

his brother Antonio to join him on Frisland. Subsequently, under the employed engagement of Prince Zichmni, the two brothers set out on explorations and military raids to the surrounding islands.

These exploits included a particularly adventurous journey when Antonio Zen accompanied Prince Zichmni on an expedition west and discovered a new land. According to the account published by Nicolò Zen the Younger, local fishermen had reported that Estotiland lay to the west and that this land was rich and populated by many nations, some of whom were cannibals. Indeed, Prince Zichmni's expedition encountered inhabited land and explored its shores. The account did not claim explicitly that this new land was part of America, yet this was clearly an association that the sixteenth-century readers of Nicolò Zen the Younger's account would have made, surrounded as they were by reports from the newly encountered world on the other side of the Atlantic Ocean, as well as by publicly circulated stereotypes such as the image of the American cannibal.[15] For example, such an image was published in Caspar Plautius's *Nova typis transacta navigatio* next to the image of Saint Brendan.

The account also details how Nicolò Zen the Elder in his travels came to the island Engronelant and encountered a monastery there with a church dedicated to Saint Thomas. In this monastery, monks from Norway, Sweden and Iceland lived in defiance of the cold, using the heat from a volcano to cook food, heat their houses and grow herbs in their gardens. Nicolò Zen the Younger attached a map to his publication, which showed the region the Zen brothers had traversed, and information from the map and the account was picked up by contemporary map-makers. Thus, on Ortelius's map, the 'St Thomas cenobium', or monastery, is located on the eastern part of the island 'Groenlandt', a horizontally oriented version of the land we know as Greenland.

Did these two Venetian brothers visit Greenland and America in the fourteenth century? The proposition is tantalizing, yet the answer is most likely no. The account is vivid, often including details that we find in other contemporary accounts or that lend a personal touch to the tale. However, Nicolò and Antonio Zen had, most likely, never travelled across the North Atlantic. Indeed, modern historians have been able to establish that at the time when they were supposed to have been roaming the Atlantic, both brothers appear at home in the records of Venetian affairs.[16]

Instead, it is likely that their descendant Nicolò Zen the Younger was not only the compiler but also the author of the account that Ortelius and many other map-makers used as a source regarding the geography of the North Atlantic. Nicolò Zen the Younger took part in a long tradition of publishing travel writing based on a mixture of the true, the likely and the wholly invented. He used themes that were common in other contemporary travel accounts, such as the report of cannibalism in the New World.

To make the account more believable, Nicolò Zen constructed a framing narrative of how he had found the 150-year-old letters of his ancestors and pieced them together. In fact, he told the reader that he had partly destroyed the letters when he was a small child, and that his publication now was a 'reparation to this present age, which, more than any other yet gone by, is interested in the many discoveries of new lands'. Clearly, Nicolò Zen the Younger knew that accounts of journeys to faraway places were of interest to his contemporaries. It has been suggested that his publication was, in fact, an attempt to bolster Venice's role in the explorations of the newly discovered parts of the world.[17]

Early modern map-makers like Ortelius presumably used Nicolò Zen the Younger's account because it was not entirely implausible that a Venetian ship could veer off its course and end up in the North Atlantic. In fact, there were other such accounts of shipwrecks that informed European views of the far north.[18] As a result of the plausibility of the account, and in despite of its more fictitious elements, a long debate followed about whether the account had a basis in an actual journey. Many have tried to identify which islands the Venetians might have visited. Already in 1898 one commentator noted, with some frustration, that 'no other story of travel ever published has given rise to such an amount of doubt, perplexity and misunderstanding extending over so long a period'.[19] Today, most scholars agree that the account can tell us more about the sixteenth-century interest in travel writing and geography than about Prince Zichmni or his Venetian comrades. Still, with a changing climate and possible new volcanic eruptions, new islands may appear in the seas surrounding Iceland. Who knows if the future will see the reappearance of Frisland and Droego on maps?

THE ARGONAUTS IN THE NORTH

The story of the Argonauts counts among the most famous travel accounts in the European literary tradition. One of the oldest stories passed down from Ancient Greek mythology, it describes a group of heroes who set out on the ship *Argo* in search of the Golden Fleece.[20] In the legend, the band of heroes travelled from Iolcus in Thessaly in eastern Greece and sailed through the Bosporus and the Black Sea to reach Colchis, a kingdom on the eastern shores of the Black Sea and the location of the coveted Golden Fleece. On the return journey the *Argo* was forced off course and its crew faced challenge after challenge, at times even having to drag their ship across land to reach Greece again.

Like both Saint Brendan and the Zen brothers, these Argonauts (literally 'sailors of Argo') encountered strange and dangerous situations on their journey and overcame many hardships before their return with the Golden Fleece. To the audience listening to the account, the story

31 The Swedish polymath Olof Rudbeck the Elder is depicted surrounded by the authorities of antiquity. Plato, Ptolemy and Orpheus look on as Rudbeck pulls back the surface of the globe to reveal that Sweden is actually *Deorum Insula*, 'Island of the Gods'.

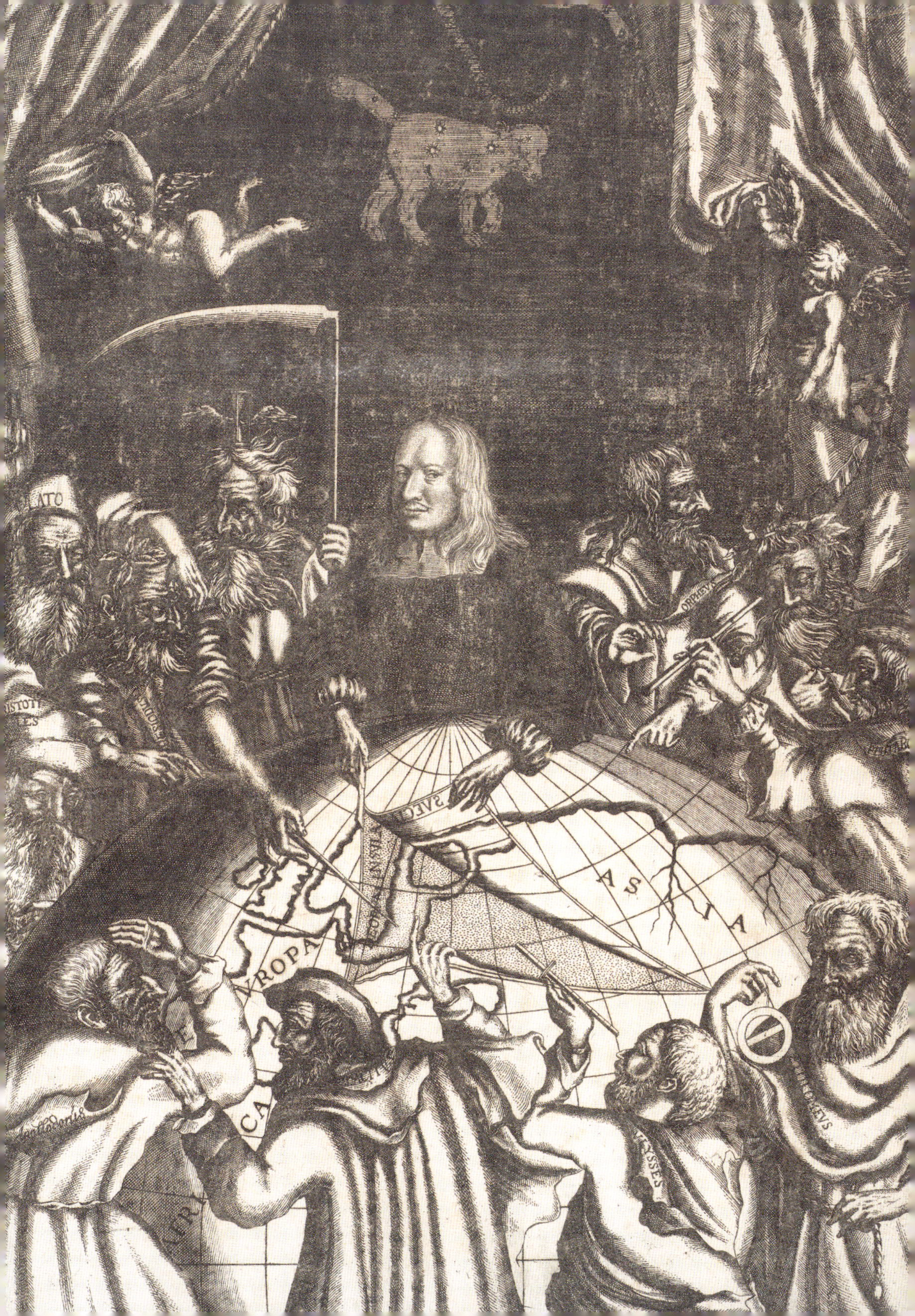
ORPHEVS
VROPA
ASIA

32 Rudbeck's map shows a northerly route of the Argonauts' return journey after retrieving the Golden Fleece. Along the coast of Norway, Rudbeck has included a depiction of the maelstrom Moskstraumen.

would have been an entertaining tale situated in a distant and mythical past, as well as an account tied to historical places, events and persons that they recognized from other stories.

There are several different versions of the tale of the Argonauts. The story has inspired generations of writers and artists, and scholars have long tried to pinpoint in more detail where the Argonauts travelled.[21] Regardless of which version one consults, though, the journey is set primarily in the Mediterranean, the Black Sea and adjoining rivers. One might ask, then, what can the Argonauts tell us about maps of the northernmost parts of the world?

If we were to ask the seventeenth-century Swedish polymath Olof Rudbeck the Elder, the answer to this question would be that the story of the Argonauts is an excellent source on the far north. Rudbeck postulated that the Argonauts, after retrieving the Golden Fleece, had eventually sailed north from the Black Sea on the River Don. After that, they dragged their ship to the River Volga and followed it until they reached the Baltic Sea at present-day St Petersburg. Once in the Baltic Sea, they continued north and found a way to the White Sea across Finland. Then the Argonauts circumnavigated all of Scandinavia and Britain (criss-crossing between imaginary islands in the Irish Sea), sailed south along the shores of France and the Iberian Peninsula, and entered the Mediterranean Sea again through the Strait of Gibraltar.[22]

To convince any doubting readers, Rudbeck made a map showing the Argonauts' route (FIG. 32). He also described in detail how the account of the Argonauts aligned with the geography of the northern regions of Europe better than with any description of the Mediterranean Sea. Here we have an account from classical antiquity, shrouded in myth, and then reinterpreted by a scholar in the seventeenth century with a considerable degree of creative licence. As such, it provides an opportunity to look closer at how mythical journeys, maps and history have been combined to propose new information about the north.[23]

Olof Rudbeck was a professor of medicine at the University of Uppsala in Sweden. He had gained international renown already in his twenties when, in 1653, he became one of the first scholars to correctly describe and identify the functions of the lymphatic system.[24] From medical studies, Rudbeck branched out to cultivate an interest in geography, philology and history. It was when he was making a map for his friend and colleague Olof Verelius that Rudbeck was struck by how similar some old Swedish place names were to names from the classical epics of Greece and Rome. He explained that 'like in a dream I felt like I had read [the Swedish place names] in the writings of the old Greeks'.[25] In fact, Rudbeck concluded that the classical Greek literature described Sweden's geography rather than that of Greece.

Out of this realization grew the *Atlantica*, a four-volume work where Rudbeck explored how different historical and mythical civilizations could be connected to Sweden. He is perhaps most famous for his hypothesis that Sweden was the lost continent Atlantis that Plato wrote about in his treatises *Timaeas* and *Critias*. Unsurprisingly, this idea appealed to the late-seventeenth-century ruling elite in Sweden, who were trying to position the country as a military, political and cultural power in northern Europe. Others were more critical of at least some of Rudbeck's claims, although the idea of connecting a country's history to ancient tradition was not alien to contemporaries; nor was the move to use maps as scholarly proof.

Rudbeck devoted a whole chapter in the first volume of his *Atlantica*, published in 1679, to explain the Argonauts' travels. To help his case he built five large ships and tried different methods of carrying and dragging them between places, concluding that 'in one and half an hour, fifty men dragged the yacht 2,400 feet, and thus eighty Argonauts could well make one German mile per day'.[26] With his calculations, this would make the northern route that he had proposed for the Argonauts the most likely.

To advance his case, Rudbeck also pointed to a map that first appeared in Abraham Ortelius's historical atlas *Parergon*.[27] In fact, this map presented the conventional interpretation of the Argonauts' journey, centring on the Mediterranean and the Black Sea. However, the map does include a small inset map showing all of Europe, and it was this detail that Rudbeck paid most attention to (FIG. 33). Rudbeck erroneously attributed the authorship of the map to the German antiquary Georgius Hornius, and lauded Hornius for his work, explaining that the error of not seeing how the Argonauts had actually reached the Baltic Sea was understandable 'since Hornius was not a Swede who [had] knowledge about the places of our country, which Orpheus [one of the Argonauts], who himself has travelled through them, places in the same order as they still exist today'.[28] In this passage, Rudbeck makes use of the importance contemporaries gave to eyewitness accounts for verifying information, yet he reinterprets the evidence. Similarly, he used only the parts of Ortelius's map that supported his own hypothesis.

This relocation of the Argonauts to the north certainly appears far-fetched today, yet the idea was picked up and discussed by Olof Rudbeck's contemporaries across Europe. For instance, the Bodleian Library invested in a copy of the *Atlantica*, and Rudbeck's research inspired several late-seventeenth-century Oxford scholars in their antiquarian pursuits.[29] The German mathematician and philosopher Gottfried Wilhelm Leibniz commented that Rudbeck had made important contributions to scholarship, although he dryly noted that these contributions had often sprung from Rudbeck's fantasy.

When interpreting the Argonauts' travel account, Rudbeck utilized contemporary conventions on making and using maps. The fact that other map-makers had taken information from the accounts of travellers like Saint Brendan and the Zen brothers to gain information about the geography of the north made Rudbeck's maps and analysis seem more plausible. To Rudbeck's contemporaries, both maps and travel accounts could be authoritative sources on geography. This enabled Rudbeck to create a feedback loop whereby his interpretation of maps and travel accounts supported each other.

33 OVERLEAF Ortelius's map of the heroic travels of the Argonauts centres on the Mediterranean Sea and the Black Sea, but a small inset map in the top left-hand corner shows a view of all of Europe, opening up the possibility of a northerly route for the Argonauts.

ARGON[AUTICA]

(inset, upper left): Argonautarum reditum et Orpheo qua tabula non capiebat, en hic eum sparsim. ATLANTICVM PELAGVS. Ierna. EVROPA. CRONIVM Mare. MACROBII. CIMMERII. HYPERBOREI. SCYTHAE. Immensa sylua. Hercules columna.

EV-RO-PAE PA[RS]

Danubius fluuius. Talaurius campus. Ipse paznifero qui non magni Isterioris. Ister Argonautarum vestigia... Alpes montes. Tergeste. CELTAE. Rhodanus flu. LIGVRES. Eridanus flu. CRO-NIVM MARE. Caucasij scopuli. NESTAEI. ENCHELEAE. Cadmi Harmoniéq monumentum. Aethalia insula. LATIVM. TYRRHENIA. Caieta. Oricum. Ceraunij montes. Stoechades insulae quae et Liguslides. OCCIDENS. Cyrnus. SARDOVM PELAGVS. Arae insula et portus, ubi Cyrces habitaculum. AVSONIA. Cerunius que et Macris, et Drepane insula, in qua Medeae antrum. DOLOPES. Olympus. THESSALIA. Pagasus sinus. Locri. ACHAIA. CVRETI. Sardo. Anthemoessa insula, ubi Syrenes. TYRRHENVM AEQVOR. PELOP[ONESI] REGIC. Alpheus fl. *Cum Imp. Reg. et Belgij priuilegio decennali 1598.* Hephesti insulae quae Planctae. Lilybaeum prom. Solis tauri pas[cua]. TRINACRIA. IONIVM MARE.

LIBYSTICVM MARE

ILLVSTRISSIMO
PRINCIPI CAROLO
COMITI ARENBERGIO,
BARONI SEPTIMONTII,
DOMINO MIRVARTII,
EQVITI AVREI VEL-
LERIS, ETC.
ABRAH. ORTELIVS
DEDICAB. L.M.

Ex conatibus Geographicis Abrah. Ortelij Antuerp.

Phila insula. Trinus fl. Syrtes siue mare vadosum ac arenosum. Myrtosium prom. Ar[—] Lotos fl. Tritonis palus. Hesperi[dum] SACER Campus.

LIBYAE

Map: Pontus Euxinus and surroundings

Main map labels

TICA.

SAVROMATAE.
Immensa silva
GETAE ARIMASTHAE
SCYTHAE
MAEOTAE
Tanais flu. qui Asiam ab Europa disterminans
ARSOPAE

PARS.

BASTARNAE
GELONES
MAEOTIS PALVS
MAEOTAE
CECRYPHAE

ALANI
Mirace
Hylea
GRAVCENII
Ophiusa
SIGYNI
Caucasus mons. Cubile Promethei.
Byces lacus
TAVRICA
SINTI
CHARANDAEI
CERCETII
BISALTAE
GYMNI
HENIOCHI
Dioscurias
BVONOMAE
Laurius campus
Cauliacus scopuli
Apollonias et Androis Flacos
Pocantis insula
TITANIA Regio
SINDI
COL
Martis et Phryxi templum
CORALLI
Salmydessus
Byzantium
AXENVM Aequor, Iasonio pulsatum remige primum.
CAVCASEVM MARE
Aeae insula
Phasis flu.
Phasis
BYZE
RAE
SAVRI
Circeus campus
CHIS
Cytaica regio

THRACIA
PROPONTIS
Cyzicum
Erythia
Sinope
ASSYRII
PAPHLAGONIA
CHALYBES
TIBARENIA
Hieros mons
MACRONES
BECHIRI
MOSCHI
DOLIONES
Scylaceon
Ida mons
Ilium sive Troia
MAVRI
MARIANDYNI
CAPPADOCIA
AMAZONES
Doeantis, sive Boeantis campus
MOSINOECI

SINTH
Sigeum
ASIAE
BEBRYCIA

EGÆVM Pelagus.
Samos
Inos Melantie petra
Sporades insulae
Thira Hipperris, et Anaphe
Rhodus
Carpathus
DICTAEVM Mare
CRETA
Minervae fanum

CYRENE

Ager.
PARS.

Inset 1 (upper right): Minoium pelagus

Amyros flu.
Euryminae
Canastreum prom.
THESSALIA
Homole
Minoium pelagus
Olympus mons
Dia
DOLOPES
Methonum
Aphete
Peneus flu.
MAGNESIA
Ossa mons
Pirasia
Anaurus et Iunonis flu. et amnium
PELASGIA
Iolcos
Ormus
AEMONIA
Pagasa
Pellace portus, et clim palus
LOCRI
Demetrias
Sepias prom.
Othrys mons

Inset 2 (lower right): Propontis / Thraciae Pars

THRACIAE PARS
PROPONTIS
BEBRYCIA
Pontus Phryxeum aequor, et Athamantidis fluenta
Posidium
Arganthonis, sive Argaius mons
Arcton mons
Cius et Olnos fines
Proconesus
Aretaeus fons
Cyzicum
CIANEA Regio
Aegeonis prom.
Cius
Abydus
Cyzicum, sive Meletia
Priepus et Pytheja
Thracicus portus
Scylaceum
Nepteius sive Halides silva
Perceote
Chitus portus
CHERONESVS
Dardania
DOLIONES
Rhoeteum
MYSIA
Ilus mons
BITHYNIA
Sigeum
Ilium
Rhyndacus flu.

Duorum horum tractuum pleniorem descriptionem quia Tabulae angustia non admittebat, hic seorsum eos delineare operaepretium duximus.

THE RIVERS OF THE NORTH POLE

One of the most elusive accounts of travel in the north is a report entitled *Inventio Fortunata* ('The Fortunate Discovery') that circulated among European scholars and map-makers for a brief period in the fifteenth and sixteenth centuries, but that has been missing since then. The account detailed the purported eyewitness description of a journey to the North Pole, undertaken in 1360 by an English Minorite friar.[30] For a time, this account was seen as an authoritative report on the geography of the North Pole and made it onto some of the most influential maps of the period.

One of the people who claimed the *Inventio Fortunata* as a source for map-making was the Flemish map-maker Gerhard Mercator. He used information from the account to make an inset map of the North Pole, included on his famous broadsheet world map from 1569. This inset map was then combined with additional sources and made into a separate map – the first making the North Pole its main subject – included in Mercator's posthumously published atlas of 1595.[31] The resulting portrayal of the North Pole (FIG. 34) is not a depiction of a barren landscape, of frozen icebergs, or of drifts of snow. Instead, Mercator marks out the geographic North Pole itself as a mountain of magnetic rock surrounded by a forceful whirlpool drawing water into the Earth.[32] Mercator writes that at the Pole 'water rushes round and descends into the Earth just as if one were pouring it through a filter funnel'.[33] Four mountainous land masses crossed by nineteen channels, or rivers, encircle the whirlpool sea. Because of the force of the whirlpool, 'no wind is strong enough to bring vessels back again once they have entered [via the rivers]'.[34] The author of *Inventio Fortunata* claimed that the water at the North Pole never freezes, partly because of the strength of the whirlpool currents, and partly because the sun shines for six months of the year at this high latitude. The land masses surrounding the whirlpool are described as mostly uninhabited, with the exception of references to people only 4 feet tall living on one of them.

The methods used for making maps pushed map-makers to seek information about the North Pole and include it on their maps. Mercator's novel map projection enables a user to plot a route between two locations on the map using a simple compass, which has made it an important tool for navigation. At the same time, the map projection distorts places far away from the equator, making the north and south look much larger than they actually are.[35] To get around this difficulty, Mercator included inset maps of the North and South Poles on his 1569 world map. Similarly, the format of globes prompted globe-makers to present information about the northern- and southernmost parts of the world. When information was lacking, the map- or globe-maker could leave blanks, add decorations or rely on the information at hand. All three strategies were used on early modern maps and globes. For Mercator and his map-making contemporaries, the *Inventio Fortunata* was useful in that it provided information about a region that, so to speak, was on their maps, yet which they knew little about. This portrayal of the North Pole from *Inventio Fortunata* appears on a number of influential early modern globes and maps. For example, the German merchant and map-maker Martin Behaim included it on his 1492 globe, and a few years later the Dutch map-maker Johannes Ruysch made it part of his influential world map.[36]

Incidentally, the whirlpool at the North Pole is not the only strong current that appears on the northern parts of early modern maps. On his world map, Mercator also makes note of a 'Maelstrom' next to northern Norway, referencing Moskstraumen, one of the world's strongest tidal currents, appearing close to the Lofoten archipelago.

34 OVERLEAF Mercator's 1595 map of the North Pole depicts four Arctic land masses separated by nineteen rivers and surrounding an open polar sea. At the top of the map, America and Asia are separated by the Strait of Anian, a feature and name from Marco Polo's thirteenth-century travel account.

Frisßlant insula

C. Spagia
Blaci
Caharu
Ilofo
Sadero gol
Ledew
Deria
Sanestol
Banar
Bondendea
portu
Campa
Anule
fort
Duilo
Rane
Ibini
Roura
Fessant
Godmer
Ocibar
Venal
Sorani
Serene
Spruie
Monaco
Ionses
Iause

Scetland insulæ

S. Magni
Scauwey
Fawle
Pressehoue
S. Bartholomæi pont
Hanglip
Swinehorne head
Faire Fl id est pulchra insula

E PARS

Obila flu.
Canogia
Zubilaga
Obila flu.
Chiagia

Californi sola fama nota

Cogib. flu.

Lago de Conibas

Hic mare est dulcium aquarum, cuius terminum ignorari Canadenses ex relatu Saguena tensium aiunt

A C A E M PARS

Gradus 75 latitudinis

Oceanus jo ossuulo prorumpit ex quibus indese septentrionem viscera terre Rupes que sub circiter 53 leucas

Hec insula optima est et saluberrima totius septentrionis

Hic europa greditur q et quorum 5 circiter m congelatus latitudinem 37 leucarum

Groclant

Mare gla.

E. Cumberlands Isles
E. Warwick Forland
L. Lumleys Inlet

Mit als Sandersons Hope prom

Na. prom
Niven pro

A furious over fall

GROENLAND

Deer flu.
Sere flu.

Berry S.
Lackers land
Wargatte

Gust prom
Wiesars mo.
Adelwik
Nax
Chax flu

OCEAN

Regne Elizabet Oft prom.

Hales iland

Grims ey
Grenastart

Cabaru
Bonden
Hofo
Ledeue
C. Spagna
Aqua
Duilo
Campa
Corps va glasker
Kroningeraerg
ISLAND

Ockhon
Frit Rane Iand
Godmer
Selsugg
Westman na
Rackenkap

Sorand
Neomi

Fodalida

TRIO

Hirta
SCOTIA

Parre insula

SEP-
TENTRIO-
NALIVM
Terrarum de-
scriptio.

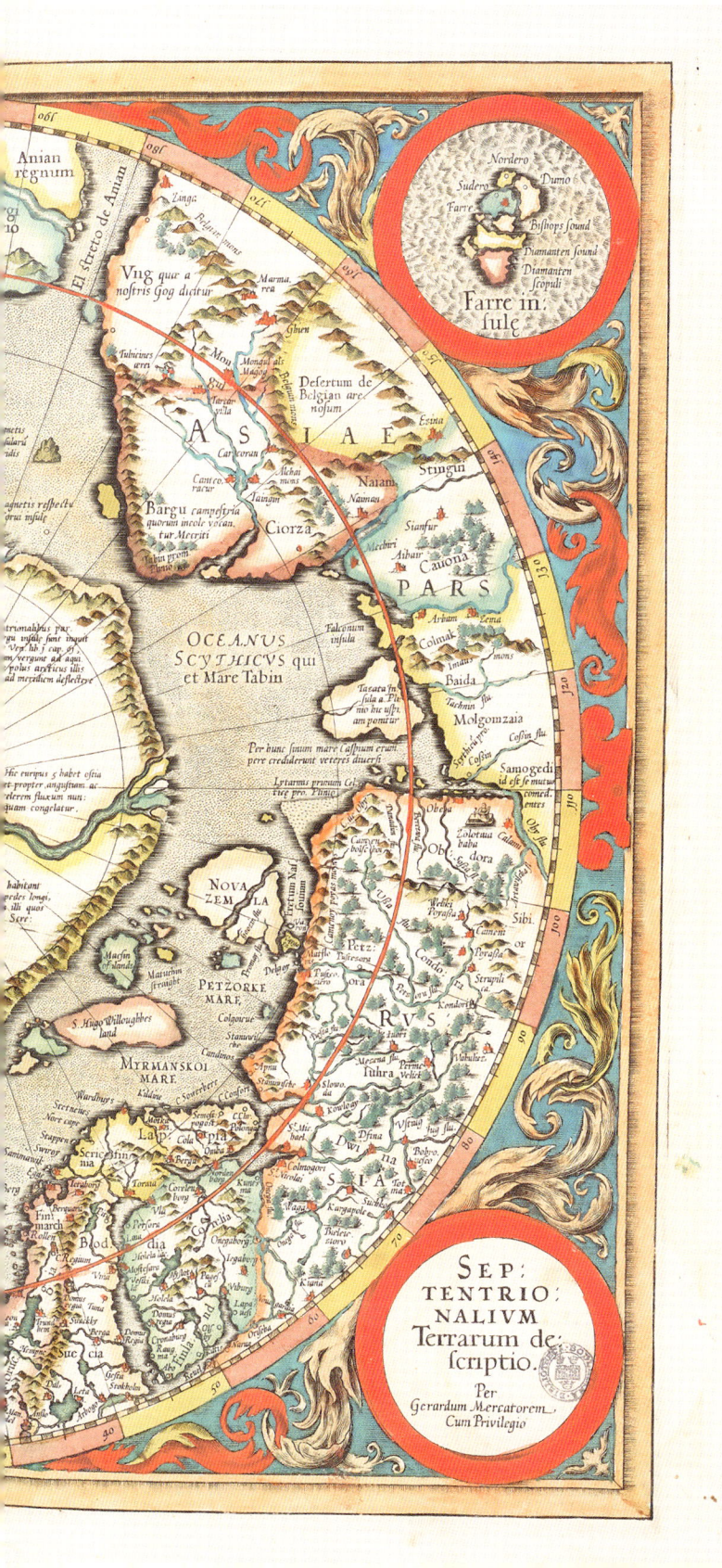

Per
Gerardum Mercatorem,
Cum Privilegio

The early Dutch word for Moskstraumen was *maalstroom*, and from that comes the English word denoting both a strong whirlpool and a chaotic situation.[37] Like Mercator, Olaus Magnus made note of Moskstraumen, but he gave the phenomenon added emphasis by drawing a decidedly outsized whirlpool on his map, and warning that 'here is the horrendous [sea monster] Charybdis'. In so doing, Olaus Magnus emphasized the exotic dangers of the north, but he also connected it to the European classical heritage with a reference to Charybdis, a sea monster from Greek mythology. When expounding on this point a few decades later, he went one step further and noted that in the case of Moskstraumen the marvels of northern nature in fact surpassed ancient myth.[38]

One might have thought that for states and merchant companies with an interest in finding a viable northern route to China, the maps portraying the whirlpool sea at the North Pole would have seemed discouraging. Though Mercator's rendition of the North Pole region does not foreground a cold climate, the whirlpool, mountains and strong currents might have indicated that finding a passage to Asia via the rivers of the North Pole was near impossible. However, it seems that this cartographical presentation did not discourage exploration. Art historian Elizabeth Sutton notes that the Dutch merchant Balthasar Moucheron studied these maps and had a discussion with Mercator about the possibility of sailing north to Asia, and that this did not diminish Moucheron's enthusiasm for the possibility of the Northwest Passage.[39] It is likely that the wealth in store for those who could find a new route to the Asian markets put a positive spin on any accounts of risks in sailing north.

At times it was even possible to interpret potential hazards as good omens, as the sixteenth-century English explorer and soldier Humphrey Gilbert did when he claimed that the strong currents of the northern seas were proof of a Northwest Passage to Asia. He also noted that he had concluded that America was an island after having conferred 'with the Mappes and Globes both Antique and Moderne'. Gilbert's proposition of a Northwest Passage seemed like a plausible proposition to the English authorities, who were tempted by the prospect of finding a new and quicker route to China that would facilitate trade. Several expeditions were sent north-west across the Atlantic Ocean, led by adventurers like Gilbert himself, Walter Raleigh and Martin Frobisher. In the end, though there is no whirlpool sea at the North Pole, the difficult geography and climate of the Arctic region meant that the expeditions failed in establishing quicker and more profitable routes to East Asia. However, what these early explorers did succeed in was providing information for more detailed maps of the North Atlantic Ocean and laying the foundation for an enduring fascination among European states, merchants and adventurers with finding the Northwest or Northeast Passage to Asia.[40]

One of the scholars and map-makers who criticized the idea of four land masses and a whirlpool sea located at the North Pole was Olof Rudbeck, who noted that 'one finds from the journeys made by us [the Swedes], the English, the Dutch and others that no one has yet been able to come right under the Northern circle of latitude because of the ice, nor there found such rivers or islands, hence one can clearly see that this [idea] is invented'.[41] This comment might seem somewhat surprising given Rudbeck's record of embracing far-fetched claims. Yet it is an example of how early modern scholars and map-makers, Rudbeck among them, evaluated the sources they used, giving primacy to what was seen as trustworthy first-hand accounts when possible. In this case, recent explorers were understood to be more reliable than *Inventio Fortunata*. That said, there were also eyewitness accounts that supported some of the features proposed by *Inventio Fortunata*. For example, the English explorer Martin Frobisher, who made three expeditions to locate a north-western passage with the aim of bringing back riches from the newly found land, described how the ships in his fleet were pushed around by strong currents.[42]

While expressing doubt as to the veracity of the *Inventio Fortunata*, the seventeenth-century German Jesuit polymath Athanasius Kircher embraced the idea that a whirlpool drew water into the Earth at the North Pole (FIG. 35). He also proposed that a corresponding spring poured it out again at the South Pole.[43] As evidence, Kircher supplied the argument that if there had been no circulation of water through a connection between the Poles via the interior of the Earth, then the polar seas would have been perpetually frozen.

In his richly illustrated treatise *Mundus subterraneus*, Kircher explains how the subterranean Earth was criss-crossed by cavities filled with air, water and fire, and that it was the interaction between these elements that gave rise to observable natural phenomena such as volcanoes, whirlpools and ocean currents.[44] He presented his readers with numerous experiments to prove his point, but he also relied on eyewitness accounts – including his own – of volcanic eruptions. That said, historian Mark A. Waddell notes that Kircher in *Mundus subterraneus* presented a view of what he proposed the interior of the Earth *could* look like (FIG. 36), rather than claim this was the exact composition of the subterraneous world. Waddell suggests that, in line with contemporary Jesuit ideals, Kircher's ambition was to prompt his readers to reflect upon God's creation by presenting likely propositions regarding what the unknown parts of the world might look like.[45] This is a useful reminder that scholarship and religion were not separate spheres in the early modern period; an investigation of

35 The Jesuit Scholar Athanasius Kircher postulated that a whirlpool at the North Pole (*top*) sucked in water that was then transported through channels inside the Earth to the South Pole (*bottom*), where the water emerged again.

170 MUNDI SUBTERRANEI

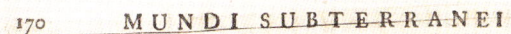

Poli Arctici Constitutio

Poli Antarctici Constitutio

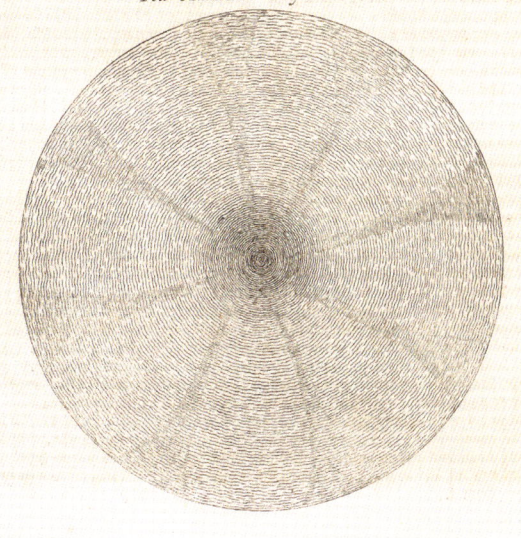

one sphere could also be construed as a contemplation upon the other.

While the idea of a North Pole region consisting of four land masses divided by rivers disappeared from most European maps in the seventeenth century, other features proposed by *Inventio Fortunata* continued to have relevance for the portrayals of the far north well into the nineteenth century. In particular, the idea of an open sea at the North Pole continued to feature both on maps and in written accounts. A remnant of this is seen in Mary Shelley's novel *Frankenstein; or, The Modern Prometheus* (1818). At the opening of the book, the narrator of the story, an Arctic explorer named Robert Walton, describes in a letter to his sister what he longs to find at the North Pole:

36 This striking view of the interior of the Earth appeared in Athanasius Kircher's *Mundus subterraneus* (1668). The map illustrates Kircher's hypothesis about what the interior of the Earth might look like, with a 'central fire' and water flowing through subterranean rivers and lakes.

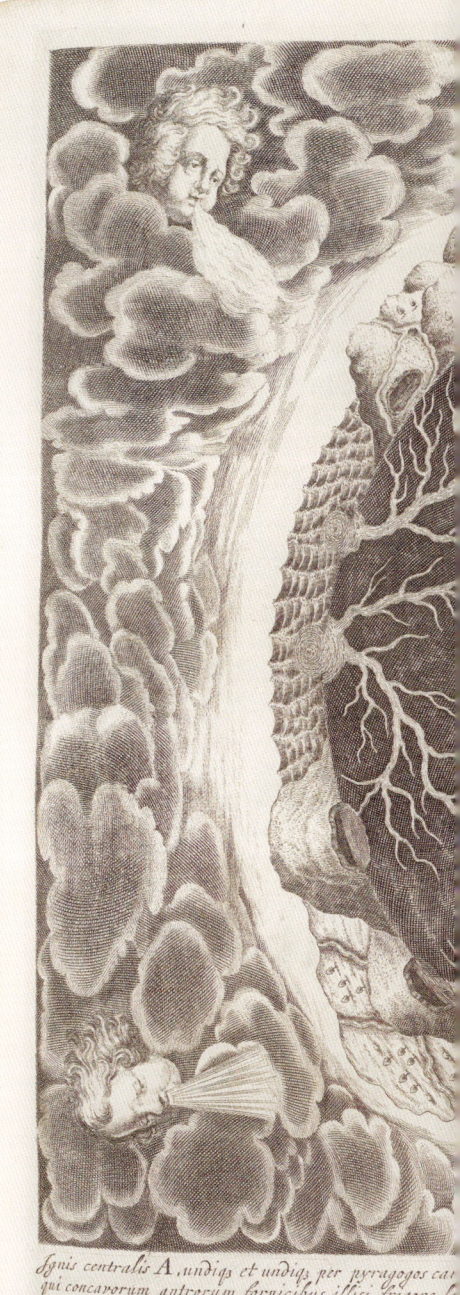

100 MAPPING THE NORTH

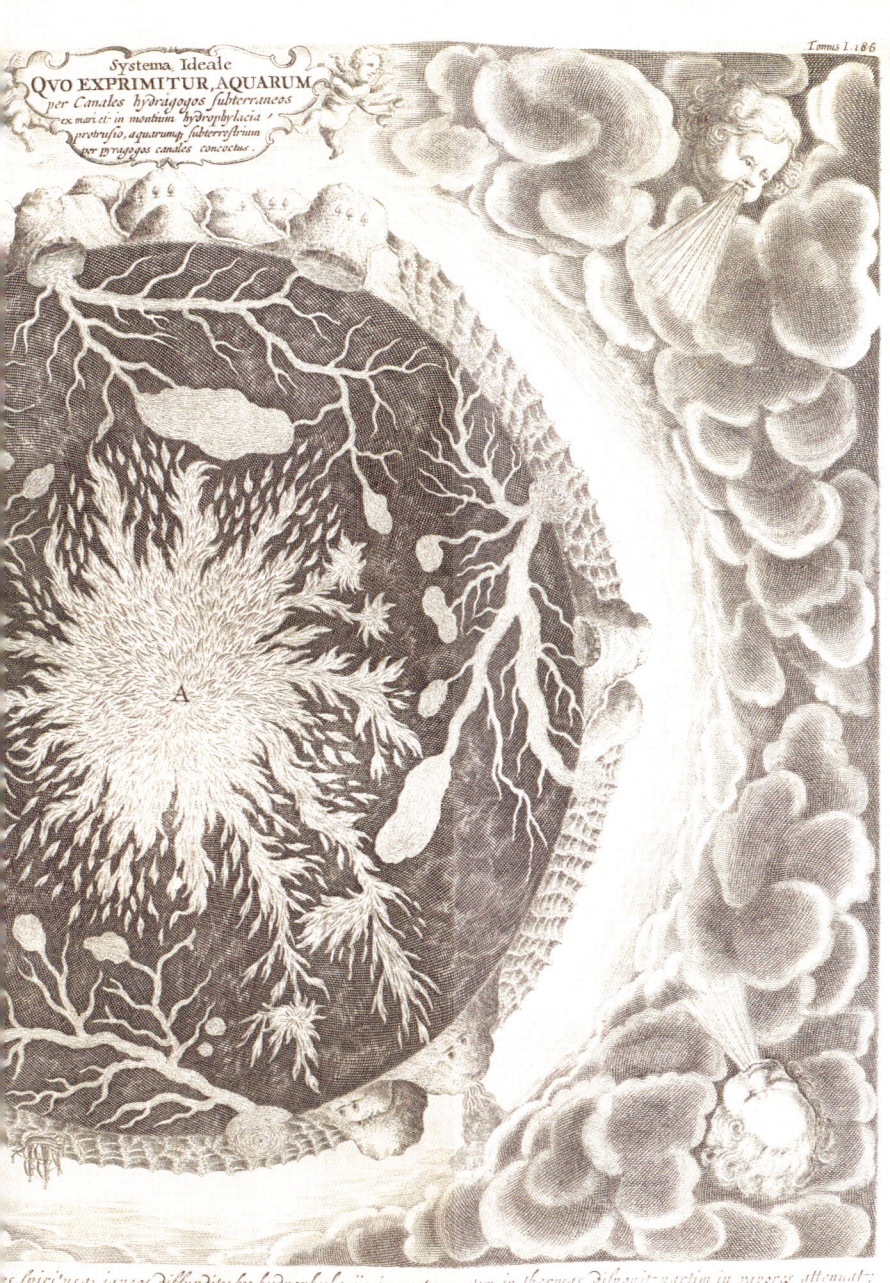

Systema Ideale
QVO EXPRIMITUR, AQVARVM
per Canales hydragogos subterraneos
ex mari et in montium hydrophylacia
protrusio, aquarumq; subterrestrium
per hydragogos canales concoctus.

s spiritus qs igneos diffundit; hos hydrophylacijs impactos, partim in thermas disponit; partim in vapores attenuat;
iquas denicq; resoluti fontes rivosq; generant; partim in alias diversorum mineralium succis fœtas matrices
materiæ fœtarum ad ignis nutrimentum destinantur. Vides hic quoq; quomodo Mare ventis et aeris præssu
am hydrophylacia ejaculetur. Sed Figura te melius docebit omnia, quam ego fusioribus verbis non explicarim.
bsequi, et hæc aërem, uti schema docet. Reliqua exactius ex ipsa operis descriptione et ratiocinio patebunt.

> I try in vain to be persuaded that the pole is the seat of frost and desolation; it ever presents itself to my imagination as the region of beauty and delight … the sun is for ever visible; its broad disk just skirting the horizon, and diffusing perpetual splendour. There – for with your leave, my sister, I will put some trust in preceding navigators – there snow and frost are banished; and, sailing over a calm sea, we may be wafted to a land surpassing in wonders and in beauty every region hitherto discovered on the habitable globe.[46]

In this passage, Shelley draws on the contemporary concern with the credibility of travel writing, making Walton admit that he puts 'some trust in preceding navigators'. She also portrays the north that Walton imagines as a serene, beautiful place, effectively creating a contrast to the brutal story about Dr Frankenstein's monster which is about to unfold. This is a slightly different view of the north to that conveyed in accounts that emphasized the hardships of travel through the north, yet both variations play into exoticization by portraying a landscape wonderous through its difference from the rest of the world.

BETWEEN MYTH AND MAP

When trying to map the north, early map-makers relied on a combination of informed conjecture, hearsay and the reports of purported travellers to the north. This means that a history of their maps of the north unavoidably becomes a study of what knowledge was considered credible at different points in time, and of what information was considered relevant to put on a map. Nowhere is this more apparent than when looking at travel accounts of journeys that never took place and how they were treated by map-makers.

It might be tempting to dismiss Kircher or Zen, or Rudbeck for that matter, as simply having made fanciful portrayals of a faraway land. However, we can better understand the societies they worked in by taking these visions of the far north as propositions that were to some degree plausible, both to their makers and to the people who viewed their maps. For example, though both Kircher and Rudbeck were criticized by contemporaries, their work was also widely cited with approval and inspired further research. The standards for legitimating knowledge have changed many times since then, yet our fascination for fictional travels remains. In many ways, modern attempts to recreate the itineraries of travellers like Saint Brendan or the Zen brothers mirror Rudbeck's interpretation of the Argonauts' adventures.[47]

Ideas about what the north is have been expressed through travel accounts and through maps, and through many other media for that matter. At times, there is no point in distinguishing between how textual accounts and maps respectively portray the north; yet in other instances the medium really did matter. In early modern Europe, the inclusion of information from a travel account on a map most likely meant that the information was presented as credible, or at least plausible. As such, when information from a fictional journey was included on a map, it gained, if only for a time, a veneer of truth. Moreover, because map-makers copied each other, the inclusion of information on one map tended to mean a wider audience and more longevity than what would otherwise have been the case. Finally, the spatial nature of maps ensures that the information added to a map ties it to specific places. We might say in this case that the map anchors the information in the north, as when St Brendan went from travelling the open seas to becoming an island. In this regard, the standards for determining truthful knowledge and the spatial nature of mapping have worked together in shaping the image of the north.

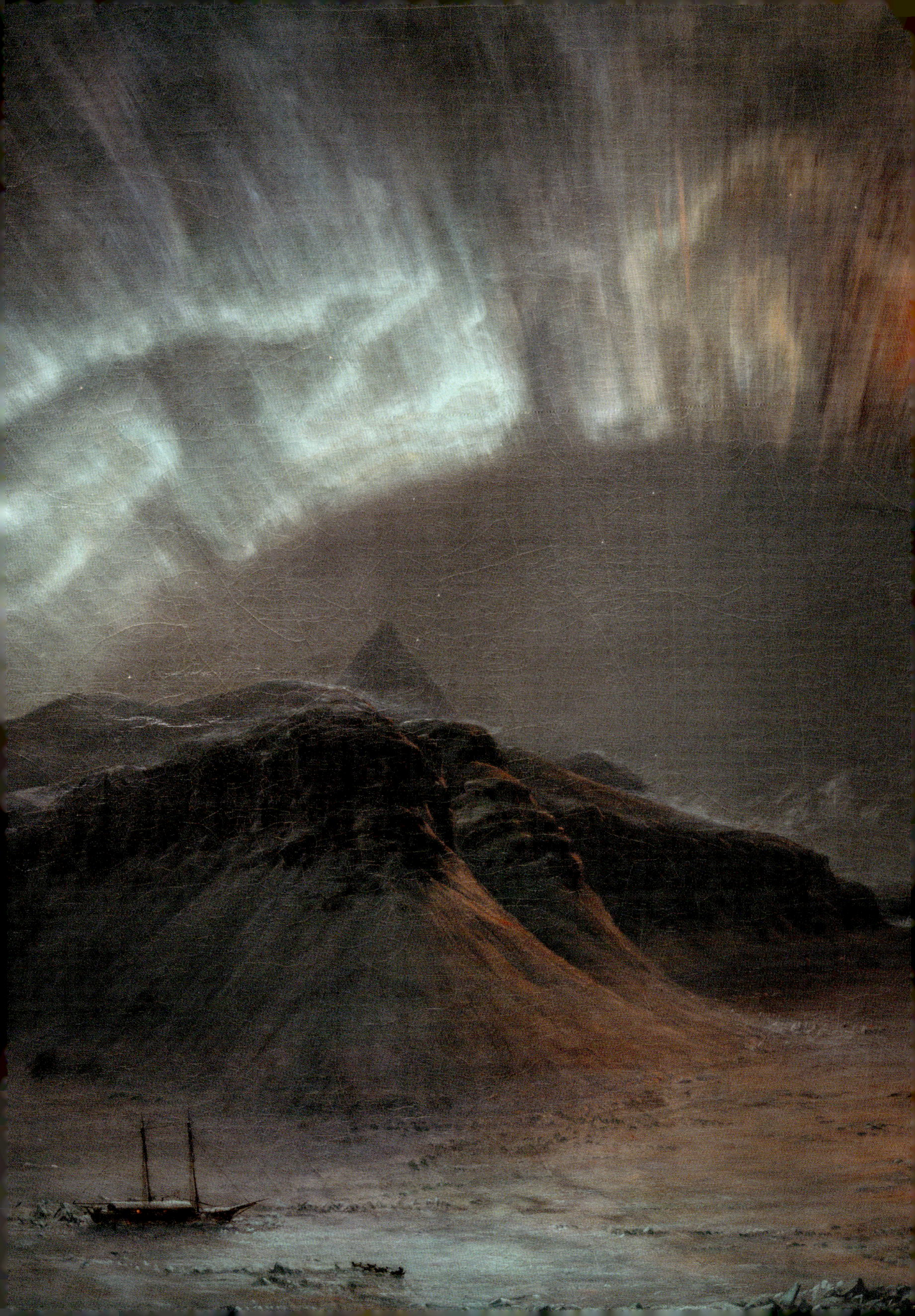

ENCOUNTERS & EXPLORATION

From the idea of an 'uninhabited frigid zone' to the paintings and prints of polar expeditions stranded in landscapes of ice, a persistent theme in descriptions of the circumpolar north has been that it is a desolate region.[1] A seventeenth-century English commentator lamented that when the ship of the Dutch navigator Willem Barentsz got stuck in the ice off Novaya Zemlya, the crew had to spend the winter 'where they had no Inhabitants, but Foxes, Bears, and Deare, to keepe them company'.[2] Echoing this sentiment, the Norwegian explorer Fridtjof Nansen in the late nineteenth century succinctly described the Arctic as 'the domain of ice'.[3] One might think that maps showing the northernmost parts of the world as a known unknown would enhance this narrative of an empty north; however, the story is more complicated than that. Instead, maps of the north have, in different ways across time and between societal contexts, promoted stereotypes about what types of people populate the north, while also obscuring the indigenous societies.

When attention is paid to how people are represented on maps, the power relations between groups and individuals which underpin

37 Frederic Edwin Church's 1865 painting *Aurora Borealis* portrays a ship frozen in the Arctic ice under a sky lit up by Northern Lights. A sleigh drawn by dogs is seen hurrying back to the ship, emphasizing the vast spaces and emptiness of the northern winter.

all mapping come to the fore. Although it is often unacknowledged in the final maps, researchers have come to recognize the central role of indigenous expertise in how maps of the north have been conceptualized and made.[4] In fact, without paying attention to encounters in the northern contact zones, it is difficult to fully understand the mapping of this part of the world. Viewing from far away and using travel accounts as knowledge are only part of the story of how the idea of the north has been formulated through maps. How people have interacted with and imagined other societies as part of the mapping process have been equally crucial.

Medieval and early modern European map-makers who resided farther south exoticized the northernmost regions of the world by populating them with half-imaginary people. As more became known about the actual inhabitants of the far north, map-makers began incorporating images of northerners on their maps and thereby reinforced the association between people and specific geographic locations and environments. Eventually, fewer depictions of people were included on European maps in the eighteenth and nineteenth centuries. In contrast, maps made by people indigenous to the north during this period portrayed rich layers of human activity. And, in a parallel development, polar explorers of the period populated maps of the north with their friends and benefactors by naming islands, straits and promontories after them, often obscuring indigenous societies while at the same time relying on indigenous expertise for making the maps. Considering these various developments, what follows is the story of how conceptions of people in the mapping process have contributed to and challenged the idea of the north over time.

MYTHS ON MAPS

Many medieval and early modern European maps include depictions of northerners to enhance the visual presentation or as a means to provide additional information.[5] The inhabitants of the north appear in many guises on these maps. This is what we would expect given the diversity of peoples living in the circumpolar north and considering that the conventions used to portray authoritative information changed over time. However, there are themes that appear again and again, and that provide evidence of pervasive ideas about people living in this part of the world.

Not least, the edges of a number of medieval European *mappae mundi* feature figures made out to resemble humans, yet set apart from humanity through aspects of their appearance. On the Hereford *mappa mundi* (see FIG. 26), made *c.* 1300, two figures with dogs' heads, so called *cynocephali*, are depicted at the northern edges of the known world, next to a group of people who are described as making clothes and saddles out of the skin of their enemies.[6] Other figures that appear along the edges of the *mappa mundi* include *blemmyes*, people without heads, or *sciapods*, who have only one leg and a large umbrella-like foot.[7] However, not all references to people living far away from the map-maker's home needed to be of a derogatory character. For example, the maker of the Hereford map has noted that on a promontory in the north-east live the 'Hyperboreans', who 'are a most blessed race, for they live without discord and grief'.[8] This idea of a hyperborean people who reside beyond the fierce north wind goes back to antiquity. Literary scholar Peter Davidson points out that people in the far north have been given such seemingly incompatible character traits through most of history: the north and its inhabitants have been seen as both barbarous and existing in a state of ideal harmony.[9] This highlights how the idea of the north is not just contained by one unified narrative, but consists of a bundle of contradictory ideas and storylines.

With some consistency, mythical people, be they *sciapods* or Hyperboreans, were located far away and beyond the possibility of any direct encounters with the map's maker or audience. Map historian Chet van Duzer has suggested that monsters, human and animal, on the margins of medieval and early modern maps were used to represent phenomena that just might be true.[10] To the medieval European, the edges of the *mappa mundi* could harbour the strange aspects of God's creation. The references on maps to semi-mythical peoples thus had little to do with the people who lived there. Some of the references were meant as illustrations of Christian theology, such as the depictions of Gog and Magog behind their wall. Other depictions drew on the classical tradition and on contemporary travel literature, replicating ideas about the wonders located just beyond the known expanses of the world.

The depiction of the north as a liminal space did not altogether preclude encounters with strange northerners. This is seen in the so-called *Vinland Sagas*, two Icelandic prose epics written down in the thirteenth century: *The Saga of the Greenlanders* and *Eirik the Red's Saga*. These accounts tell of the Norse exploration of the North Atlantic Ocean around the year 1000 and the Greenland settlements resulting from this.[11] There is a description in *Eirik the Red's Saga* of how the Norse travellers encountered a *sciapod* on an expedition along the North American coast. The *sciapod* shot one of the Norsemen, which prompted the rest to take a different route and not 'put their lives in further danger'.[12] The encounter with the *sciapod* would have brought home to the audience that what was described was a perilous journey going beyond the common routes of travel. The Icelandic Sagas build on oral accounts about the experiences of past generations, yet they were also influenced by contemporary thirteenth-century themes, such as the idea that monstrous people like the *sciapods* could be found in faraway lands.

It is worth noting that the location of those faraway and strange places changed with the perspective of the author or map-maker. On the Hereford map the Nordic region was placed at the outer,

38 Made by the Islandic bishop Guðbrandur Þorláksson, this map from 1606 shows the North Atlantic as a Norse domain. At the same time, the map also includes new geographical information, such as the fictive island Frisland described by Nicolò Zen.

Literarum quæ in hac mappa occurrunt significatio, ab ipso Dno Gudbrando annotata

A. *Snæfells Jokull mons Islandiæ Occidentalis, conspicuæ altitudinis.*
B. *Hvitserk, mons Grönlandiæ.*
C. *alter mons Grönlandiæ nivosissimus nec minoris altitudinis, verso Africum distans à Huitserk 24 dierum itinere.*
D. *tertius mons lateris Orientalis Grönlandiæ, & proximus sinui Eriksfiord. Jokullfiall skal maud suffua vest i Nord før en mand kand komme ud i Grönlandtz fiorden.*
E.E.E. *Latus orientale Grönlandiæ inhabitatum, habens sinus plurimos & promontoria, item glaciem magna copia litus illud cingentem, quæ non nisi ab Africo & Lybonoto in mare hyperboreum extruditur, alias continue adhærens cujus pertæsi Nautæ semel atque iterum iter Grönlandicum frustra tentarunt. fuit autem Norvegis olim navigatio in Grönlandiam, ex oppido Bergensi, primum versus occidentem, mox cursus versus corum nonnihil flexo usque dum promontorium Grönlandiæ Heriolfsnes assecuti fuissent.* Cnarratione.
F. *sinus Grönlandiæ maximus Eriksfiord ab inventore Islando sic dictus ut patet ex ipsa*
G. *Alter Grönlandiæ sinus sine nomine, nisi qui dicebatur Westrbijgd occidentalis habitatio.*
H.H.H. *Latus Grönlandiæ occidentale, inhabitatum & incognitum veteribus.*

semi-mythical, edges. In the *Vinland Sagas*, however, the Nordic realm was the homeland. Here, the exotic encounter occurred across the western ocean. The Hereford map and the *Vinland Sagas* use similar literary and visual symbols to portray the exotic, but they locate them in different places. Such a discrepancy would not necessarily have troubled a contemporary audience of either the Hereford map or the *Vinland Sagas*, since the exact geographical location was, most likely, less important than the conception of a liminal place far away from home that the *sciapod* invoked.

Over time, the medieval Norse settlers of Greenland, who were the heroes of the *Vinland Sagas*, themselves achieved semi-mythical status. The settlements were abandoned before the dawn of the sixteenth century. However, early modern scholars in Denmark–Norway, which at the time included Iceland, began to show a renewed interest in the Old Norse accounts, and the stories of past exploits became the focus of historic investigations with patriotic overtones from the late sixteenth century. As a part of the investigations, these early modern scholars made maps that provided the first spatial representations of the medieval Norse North Atlantic settlements (FIG. 38). They portrayed Greenland as belonging to the same geographical sphere as their Nordic homeland when they mapped out the itineraries and settlements of the Norse travellers. They were describing historical events, yet they were also making a political argument in their own time about Greenland belonging to Denmark–Norway. This reflected the ambitions of the Danish Renaissance monarchy, which wished to capitalize on the claims over Greenland emanating from the sagas. Nor was this simply a theoretical exercise; the Danish Crown sent numerous expeditions to Greenland to investigate past connections and enforce new ones.[13]

PEOPLE ON MAPS

Myth and the mundane are present to an equal degree in Olaus Magnus's *Carta Marina*. While some illustrations relate directly to practices Olaus Magnus had observed, other images drew more on classical and medieval learning and myth.[14] For example, two figures on the coast of Greenland portray a full-length man fighting with a short-statured man. This is a reference to the Greek Homeric epic *Iliad*, which mentions a fierce and short-statured people who live in the far north. Olaus Magnus elaborates many of the themes of *Carta Marina* in his monumental *Historia de gentibus septentrionalibus* ('Description of the northern peoples'), including the idea of a short-statured people living in Greenland (FIG. 39). With references to ancient authorities such as Pliny and Herodotus, he describes how 'the pygmies of Greenland' would ravage nearby crane nests each year to make sure that the bird

39 Several early modern map-makers, including Olaus Magnus, noted that so-called pygmies lived in the Arctic. These remarks did not build on actual encounters with people. Instead, the map-makers relied on Greco-Roman geographical authorities such as Pliny the Elder, combined with the habit of placing the unknown and strange at the edges of the map.

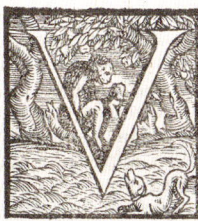

REGIONES
SVB POLO ARCTICO

Auctore Guilielmo Blaeu.

FRIGVS inges illic habitant
Et jejuna FAMES.

New South Wales.
Buttons Bay
New North Wales
Baffins Bay
James his Bay
NOVA BRITANNIA
Fretum Hudson
FRETVM DAVIS

*Amplissimo Spectatissimo
Prudentiss.° viro*
GVILIELMO BACKER
DE CORNELIIS,
Reip. Amstelodamensis Consuli et Senatori, nec non in Consessu Societatis Indicæ Orientalis Assessori Tabulam hanc D.D. Ioh. Blaeu.

Terra de Cortereali
TERRA NOVA
YSLANDIA

Circulus sub Circulo Arctico

CATHAIA
Gnduc Cambalu

A:
TARTA:
SIÆ

MARE
TARTA:
RICVM

Tazata Insula

PARS

Molgomzaia

Obd ora

NOVA ZEMLA
NIAREN MORE

Spitsberge

MOURMANSKOY MORE
Nova Hollandia

LAPONIA
EVRO
SVECIÆ PARS
PÆ PARS

RVSSIÆ PARS
SAMOIEDA

Milliaria Germanica communia
pro 60 latit. gradu
pro 70 latit gradu
pro 80 lat. gradu

population would not grow too numerous.[15] Abraham Ortelius's map of northern Europe similarly includes a note on the outline of Greenland specifying that 'Here are pygmies', as does Claudius Clavus's map of the Nordic region. On these maps, deep-rooted traditions of the mythical and semi-mythical were mixed with new geographical information.

As a general trend, the depictions of people that had been so noticeable on some medieval *mappae mundi* moved from the centre of European maps to the margins in the sixteenth and seventeenth centuries. This, however, did not make the illustrations any less powerful as conveyors of knowledge and prejudice. The decorative illustrations, known as cartouches, in the margins of maps emphasized the natural riches and cold climate of the north, but also the manner of dress and the physical characteristics of northern people. All of these elements are present on the seventeenth-century Dutch map-maker and publisher Willem Blaeu's map *Regiones Sub Polo Arctico* (FIG. 40).[16] The map's cartouche is flanked by two figures. On the left, a wild man is gnawing on a piece of meat; on the right, a personification of winter is warming his hands over a bucket of embers. In addition, the lower right-hand corner of the map shows two hunters clad in fur-lined clothing, who seemingly debate whether to attack the large bear standing next to them. Two smaller fox-like animals have opted for a safer strategy, scuttling away from the bear. The two men are not identified as representatives of a particular nation, yet their dress and the pair of skis one of them carries locate them as belonging to colder climes. In contrast to the decorative cartouches of Blaeu's other maps of European nations and regions, this map does not display any references to architecture, the arts or warfare. Instead, Blaeu has emphasized the harsh climate and the emblematic fauna of the region, together with the vaguely winter-clad hunters.

40 PREVIOUS SPREAD The possibilities of navigable waters remain alluringly open on this Dutch map of the Arctic from 1640 by Blaeu. Still, the map characterizes the Arctic as a region with a harsh climate through the depiction of a wild man gnawing on a bone next to the personification of winter trying to warm his hands.

Early modern ethnographic references were elaborated further in a particular genre of primarily Dutch maps that were popular in the seventeenth century. Sometimes called *cartes-à-figures* or 'maps with decorative borders', these consist of regional maps surrounded by cities of the region and miniatures detailing the typical costume and imagined physical features of people from different places. The ethnographic depictions on *cartes-à-figures* tapped into several early modern trends and interests. In particular, the decorative borders have clear parallels in contemporary costume books. These were commonly organized geographically, and, like the decorative map borders, they inventoried the dress and bodily characteristics associated with people from different parts of the world. The educated European reader who browsed a costume book to learn more about different societies might also have rested their eyes on a *carte-à-figures*. Like the period's lavish curiosity cabinets, the costume albums and the maps with ethnographic representations along the borders were attempts to collect, categorize and by extension control the world.[17]

One example of this is the case of the Swedish seventeenth-century nobleman Nils Bielke, who owned a rare set of four large wall maps of the continents Africa, America, Asia and Europe mounted with a set of engravings of people from different parts of the world.[18] The maps decorated the walls of Bielke's residency when he was the governor of Swedish Pomerania, 1689–1697, and they would certainly have made an impression on visitors, measuring 180 × 145 cm and displaying beautiful colours (FIG. 41). The maps conveyed information about different parts of the world, but in Bielke's study they did more than that. For their owner, the maps were status objects, as were the large library and collections of weapons and curiosities that Bielke also owned.

The four wall-map compositions that Nils Bielke displayed in his residence were compiled by the Dutch engraver and publisher Johannes De Ram, while the maps were by Frederick De Wit and Gerard Valk, mounted together with a set of engravings of people from different

ENCOUNTERS & EXPLORATION 115

NOVA TOTIUS AMERICÆ TABULA

AMERICA SEPTEN- TRIO-

OCEANUS OCCIDENTALIS

MAR DEL ZUR

AMERICA

LAMERIQUE.

parts of the world.[19] Of the forty-eight ethnographic costume illustrations on Bielke's maps, five depict places that were described as part of the north in other contemporary European maps and treatises: Canada, Strait Davis and Hudson Bay, Archangelsk, 'Laplant' and Stockholm. These images do not convey one and the same idea of life in the north. For example, the image of Strait Davis and Hudson Bay foreground the natural riches and lucrative trade opportunities of the region in the form of depictions of fish and fur-coated animals (FIG. 42). The European man and indigenous woman on the engraving wear fur-lined clothing; a large mountain can be seen in the background. On Nils Bielke's copy of the engraving, the impression of a cold climate is further emphasized through the white colouring of the background.

41 This large wall map of America by Frederick De Wit and Gerard Valk functioned as a status symbol, while also providing new information to its seventeenth-century audience. An inset map provides a view of the circumpolar north, dominated by a blank space around the North Pole.

ENCOUNTERS & EXPLORATION

42 The central theme of this late-seventeenth-century engraving of Hudson Bay from the same series as FIG. 41 is the bountiful resources of North America, exemplified through the trade in fish and furs. Mounted together with a map, the image created a stereotyped view of a place and the people who inhabited it.

As a whole, this is an engraving that emphasizes nature over human habitation. In contrast, the image of Archangelsk from the same set of engravings features a busy port with ships ready to sail. In the foreground a man and a woman richly dressed in furs and brocade provide a reference to trade links between western Europe and northern Russia that contemporary educated viewers would have recognized. Bielke's wall maps with the accompanying ethnographic prints thus presented an inventory of what people looked like in different parts of the world, but they did not project one unified image of the north.

This, however, did not mean that engravings did not also conflate ethnographic information and create stereotypical representations of people. The depiction of 'Laplant' on the map of Europe is an example of this type of conflation (FIG. 43). The term 'Laplant' was used in the early modern period to refer to northern Scandinavia and the semi-nomadic Sami who lived there. The image shows a Sami man and woman wearing winter clothing and short skis or snowshoes. In the background a reindeer is seen drawing a sleigh across the snow. The man also carries a kayak, and whales are seen in the distance. There were Sami who maintained their livelihoods from fishing and whaling, though these features are less common in contemporary illustrations of Sami life, which tended to foreground raindeer herding and pagan

43 This engraving from the same series provided early modern viewers with a stereotype of Sami appearance. In the background a reindeer is seen pulling a sleigh, a motif that also appears on Olaus Magnus's *Carta Marina*, in Johannes Schefferus's *Lapponia*, and mirrored in later travel accounts, such as that written by the nineteenth-century explorer Arthur de Capell Brooke.

ENCOUNTERS & EXPLORATION

THE HISTORY OF
LAPLAND
Wherein are shewed the
Original, Manners, Habits,
Marriages, Conjurations, &c.
of that People.
Written
by
Iohn Schefferus
Professor of Law & Rheto
rick at Upsal in Sweden.

At the Theater in Oxon 1674.

religious practices. Examples of this are seen in Olaus Magnus's *History of the Nordic Peoples* and in the influential treatise *Lapponia* (FIG. 44), written by the Swedish scholar Johannes Schefferus.[20]

Thus, the early modern maps with decorative borders combined an attention to detail and local variation with a conflation of different groups of people living in the north. On that point, historical geographer Peter R. Martin notes that ethnographic depictions often portray the indigenous peoples of the north as passive figures, or even as 'anthropological curiosities'.[21] *Cartes-à-figures* certainly present people as curiosities, though they depict societies from all parts of the world in this manner. It is rather through the accessories – hunting implements or rich silks – and surrounding structures – tents or stone mansions – that the maps create implicitly hierarchical distinctions between different peoples.

EXPLORERS CLAIMING AND NAMING THE NORTH

In 1818 the British map-maker Aaron Arrowsmith published a map entitled *Map of the Countries Round the North Pole*.[22] Versions of this map appeared throughout the nineteenth century and into the twentieth, becoming one way in which an interested British public could visualize the geography of the Arctic at a time when each new report was soon followed by another. How did Arrowsmith go about delineating his map, and what did it highlight and obscure in terms of the people living in this part of the world?

To begin with, and despite a name promising a map of the *countries* located around the North Pole, Arrowsmith decided to make a circular map that extended from the North Pole to 50°N, cutting across all national borders at this latitude. As a result, the boundaries of his map

44 First published in 1673, Johannes Schefferus's description of the Scandinavian Sami became influential in shaping European perceptions. Two centuries later, Arthur de Capell Brooke still referred to Schefferus as an authority and to his map as a useful source on the geography of northern Scandinavia.

MAP OF THE COUNTRIES ROUND THE NORTH POLE

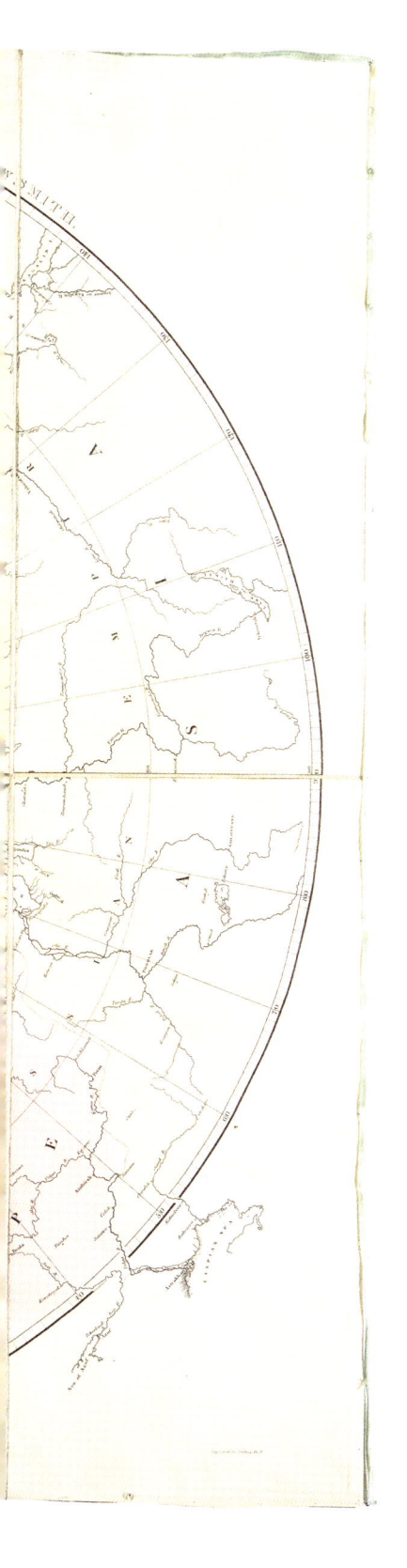

speak more to geography than to political territory. This impression is heightened in the copy reproduced in this book (FIG. 45), since its colouring (added after printing) does not follow national boundaries either. In fact, the colouring is organized along the lines of continents. Greenland, a Danish colony at the time of the map's publication, is coloured green like the rest of North America, and Russia's European part is coloured red, while its territory in Asia is yellow.

Thus, at first sight, Arrowsmith's map of the circumpolar north seems relatively devoid of power politics. However, the map is not a neutral portrayal. When Arrowsmith published his work, the northernmost reaches of North America were part of either British North America or Russia, or consisted of lands that had not been claimed by any southern country. This did not mean that this region was unpopulated. On the contrary, people indigenous to the north, organized in nations and with territorial claims, lived on the vast swathes of land that appear empty on Arrowsmith's map. By excluding this information, as well

45 Despite its name, Aaron Arrowsmith's *Map of the Countries round the North Pole* does not demarcate the boundary lines between different states located above 50° N. Still, through place names and notes about geographical discoveries, Arrowsmith mapped this as a region controlled by southern states.

ENCOUNTERS & EXPLORATION   123

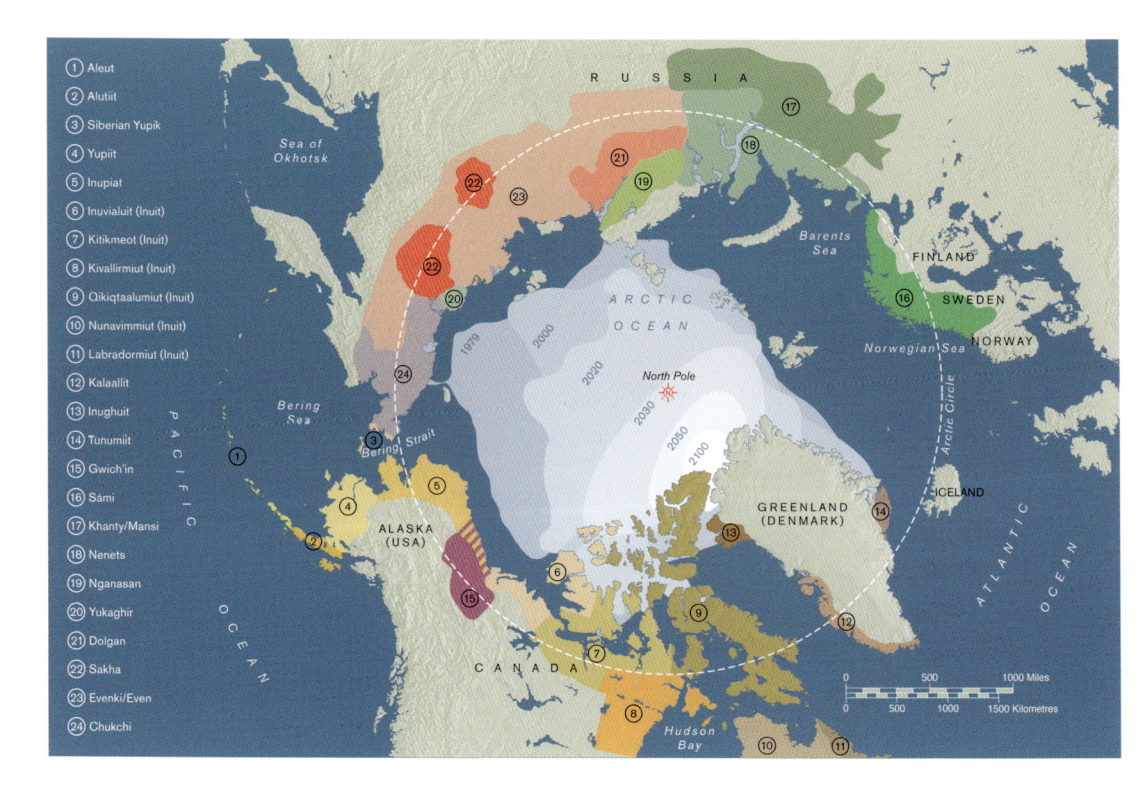

46 This modern map gives an overview of the diversity of indigenous peoples that live in the circumpolar north. Many live in cities and towns, and lead lives similar to what can be found in other parts of the world. Others adhere to ways of life that have been common in the Arctic for much longer.

as through the choices of colouring, Arrowsmith's map presented a version of geography that erased indigenous political structures.

The map is also incomplete in terms of geography, with coastlines ending abruptly and colours fading into white, reflecting the partial state of geographical knowledge about the northern latitudes in Britain in the early nineteenth century. At some places, Arrowsmith has indicated the incompleteness of information through dotted lines or through formulations such as the '*supposed* Strait of Juan de Fuca' marked out south of Vancouver Island in south-western Canada. Arrowsmith has acknowledged some of his sources through measures such as the legend in north Canada which, in the middle of a blank part of the map, evocatively reads 'The sea seen by Mr [Samuel] Hearne 1771'. Less visible are the indigenous people who provided explorers like the British naturalist and fur-trader Samuel Hearne with information.[23]

During most of the nineteenth century, Arrowsmith, and later his successors, continued to update the map to match incoming reports. From 1875 the map-maker and publisher Edward Stanford produced a North Pole map based on Arrowsmith's map, but superimposed it with the dates and destinations of a long list of 'Arctic Worthies', from Sebastian Cabot's voyage in the North Atlantic in 1497 up to the time of publication.[24] Much like the map of the Arctic published in *Stielers Hand-Atlas*, Stanford's updated version of Arrowsmith's map created a narrative of the growth of geographical knowledge and the Arctic as a region gradually discovered by southern explorers. The maps thus simultaneously presented the north as empty, and indicated that southern states had historical claims here through their presence in the region. In this, the maps were part of a nineteenth-century discourse which, as historical geographer Michael Bravo has pointed out, framed the expeditions to the North Pole and the South Pole as a coherent history of exploration, and in the process claimed the Arctic as a domain belonging to the explorers and their backers.[25]

Another way of claiming the Arctic has been to describe and propose names for geographical features and settlements.[26] A map of Svalbard by the English whaler, scholar and later clergyman William Scoresby Jr gives an interesting view into this long-standing practice (FIG. 47). The map was published as a part of Scoresby's treatise *An Account of the Arctic Regions* (1820), in which he described geography, climate and animal life in the Arctic in general, and the conditions of the whale trade around Svalbard in particular. In his description of Svalbard (using the earlier name Spitsbergen), located in the Arctic Ocean, Scoresby commented that the 'country exhibits many interesting

Surveyed & Drawn by W. Scoresby Junr.

Engd. by W. & D. Lizars Edinr.

Edinburgh. Published by Constable & Co. 1820.

views, with numerous examples of the sublime'.[27] His textual description emphasizes the desolate nature of the largely uninhabited archipelago.

In contrast, the place names included on Scoresby's map evoke connections with southerly states, people and phenomena. For example, several place names relate to the states that had been involved in the Svalbard whale trade since its beginnings in the seventeenth century. Such names included 'English Bay' (today Engelskbukta), 'Muscovy Mt' (Hedgehogfjellet) and Danes I[sland] (Danskøya). All of these names were well established by the time Scoresby made his map, but he also included some new proposals of a more personal nature. For example, his interest in theology shows in the mountain he named 'Mitre Cape' (Kapp Mitra) because he considered it 'having the form of a mitre', the headdress of a bishop. Scoresby's map contributed new knowledge about the geography of Svalbard and presented itself as a scientific survey; yet the map is, no less than Arrowsmith's, an act of promoting certain world-views over others.[28]

Place names such as English Bay and Danes Island are not just evidence of the diversity of people active in the Arctic, but also reflect competition between southern states to control these areas. In the case of Svalbard, the question of which country the islands should belong to was settled in 1920 with the Treaty of Svalbard (originally Treaty of Spitsbergen), which made it a part of Norway, though with some provisions to accommodate the claims of other nations.

The late nineteenth and early twentieth centuries saw several other conflicts over land in the Arctic. Maps played a central role in both instigating and resolving these conflicts, often at a cost of human lives in the balance. One example of this is the case of the fictitious Peary Channel in north Greenland.[29] The American explorer Robert E. Peary claimed that he had observed in 1892 how a channel separated

47 William Scoresby, the whaler-turned-scientist who became a clergyman, reveals his personal interests on this map of Svalbard, naming a mountain in the west of the archipelago 'Mitre Cape'.

Greenland from a large island to its north. In a somewhat typical display of self-aggrandisement, he named the channel after himself. To support and broadcast his observations, Peary made a map of the region. This map and his observations were used by the Denmark Expedition of 1906–08 as a basis for planning the last stages of their journey, which ended with the death of the expedition leader Ludvig Mylius-Erichsen and two of his companions. Peary had hoped that he could claim the island north of the Peary Channel for the United States. In contrast, Denmark – which had had a colony on Greenland since the eighteenth century – was anxious to establish a presence in the region and reinforce its claim over all of Greenland. Thus, there was a backdrop of international competition to both Peary's expedition and the ill-fated Denmark Expedition.

In the case of the Peary Channel, however, elucidation involved personal drama as much as high politics. When Mylius-Erichsen was reported missing, the Danish authorities commissioned the explorer Ejnar Mikkelsen to sail north on the ship *Alabama* to establish what had happened. After a series of mishaps, Mikkelsen and the ship's mechanic Iver Iversen found themselves stranded alone during not one but two long Arctic winters. The two managed to survive against all the odds, and they could finally return to Copenhagen in 1912 and report that they had discovered a cairn left by Mylius-Erichsen with a letter in which he stated that 'Peary Channel does not exist'. The letter also mentioned that Mylius-Erichsen had renamed Independence Bay 'Independence Fjord' – giving the inlet a more Danish flavour. He also named a smaller fjord 'Brønlund Fjord' after the Greenlandic Inuk explorer Jørgen Brønlund, who was part of the expedition, and another fjord 'Hagen Fjord' after the expedition map-maker Niels Peter Høeg Hagen. All three of these place names survive to the present day, though both Brønlund and Hagen perished in north Greenland together with Mylius-Erichsen.

The practice of describing and naming places could fulfil numerous functions for explorers and map-makers. It could be a way to honour the financial backers of an expedition – such as Cape Morris Jesup on Greenland, which Robert E. Peary named after his benefactor; to commemorate or extol the virtues of expedition members, friends or family, like Brønlund Fjord, or Scoresby's Sound on Greenland, the latter named by William Scoresby in honour of his father; or to promote the claims of one country over another, as in the case of both Peary's and Mylius-Erichsen's naming habits.

The act of giving names to places in the north often had the additional effect of obscuring indigenous place names. Maps give an impression of filling blanks, of bringing an empty north into the sphere of civilization through naming. However, that could not be done without also obscuring those already living in the north. The silences on the maps discussed here show starkly when they are placed next to a map of population in the circumpolar north today (FIG. 46). The places people call home now are not necessarily the same as those of their ancestors. People have moved of their own accord, but they have also been forcibly relocated by expansionist southern states and economic interests. Moreover, a changing climate could push relocation further. Still, a map of the peoples of the north gives some indication of the numerous indigenous cultures that call the northernmost reaches of the world their home, from the Sami in northern Scandinavia and Russia, via the Inuit in Greenland, Canada and Alaska, to the Yupik in Siberia and Alaska. These are nations that cross the borders of states, both historically and today.[30]

The idea of an empty north relies on a distinction between 'nature' and 'culture' that itself has a history. To the medieval and early modern European, both the human and the natural world were God's creation, and that was much more important than any nature–culture dualism.[31] In this context, the placement of *sciapods* and similar creatures on a map was less a comment on whether people lived in the north, and more a characterization of this region as unknown and exotic. Ironically, as southerners began to learn more about the people

ENCOUNTERS & EXPLORATION

who did live in the far north in the eighteenth and, increasingly, the nineteenth century, the exclusion of northerners from maps took on an explicitly political dimension in that it was tied to claims of land.

BUSY NORTH

The Bering Strait appears as a busy thoroughfare on a nineteenth-century map made by a person from the Siberian Yupik people (FIG. 48).[32] Precious little is known of the maker of the map. Most likely he or she came from the Chukchi Peninsula on the western shores of the Bering Strait, where Asia meets America. The map-maker clearly saw this as a populated region and as a place for encounters. On the water, indigenous kayaks and Western-style vessels brush past each other. In the lower right-hand corner of the map, as it is displayed here, a high-masted ship is anchored to the shore, presumably a whaling vessel settled for winter quarters. What might be a trading scene is depicted on the shoreline next to the ship. The map also includes scenes of hunting and other everyday activities, and many different kinds of animal roam the seashore and ice, or swim in the water. This is a portrayal of a busy place, in appearance a melting pot, if not in terms of climate.

The map of the Bering Strait is the only one of its kind that survives; it is possible that it was made with a view to interesting an outside audience, rather than being used by its maker. Historians have noted that while the human and animal figures on the map are reminiscent in style of other Siberian Yupik art objects – like carved bow drills (FIG. 53) – the map form itself is unusual.[33] That said, the map is certainly no simple tourist souvenir. It includes a wealth of geographical information, and it is detailed enough that

48 This nineteenth-century sealskin map of the Bering Strait was made by an anonymous person from the Siberian Yupik people. It shows the geography of the northern Pacific Ocean together with scenes of whaling and trade.

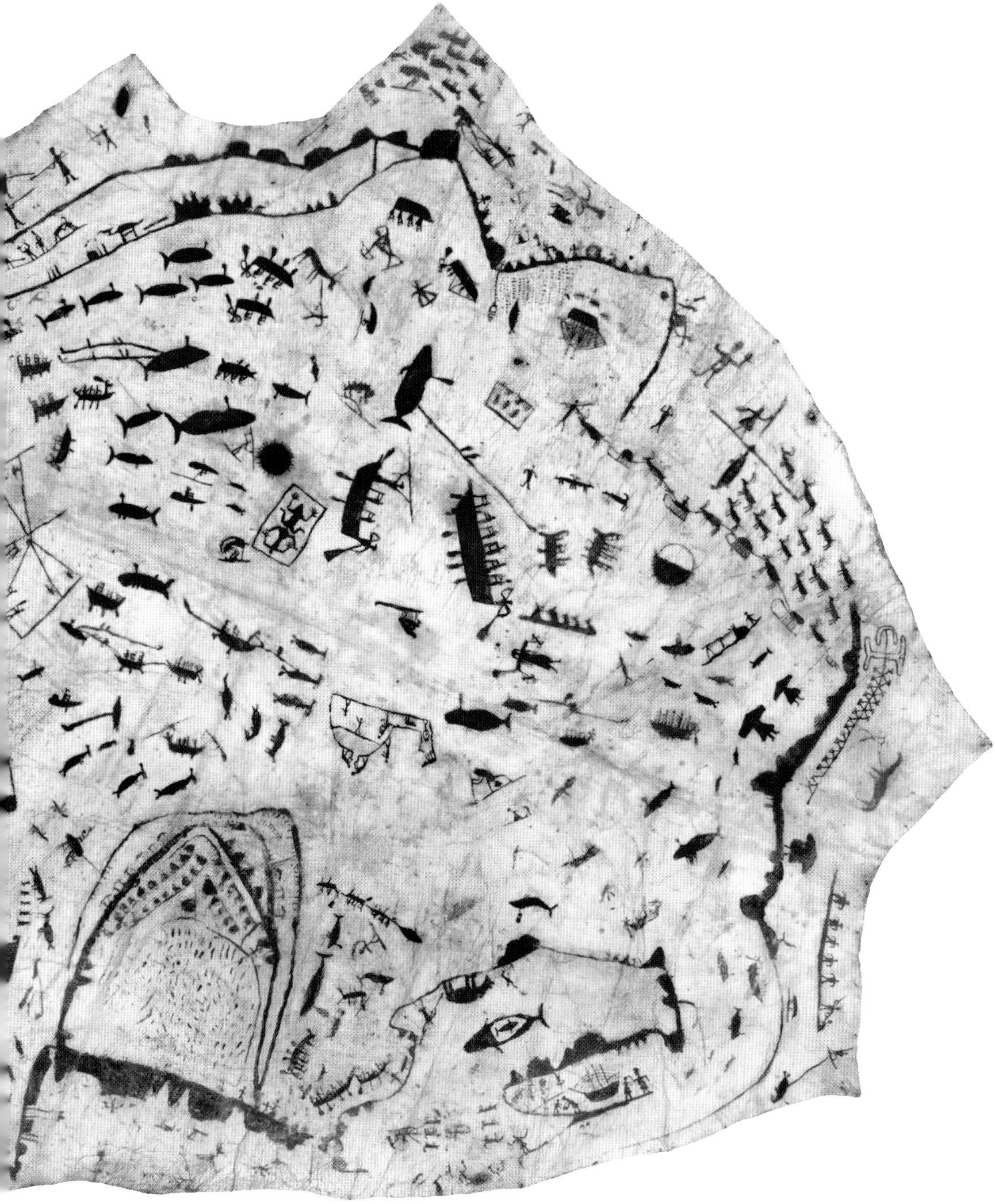

49 This set of late-nineteenth-century maps shows part of the east coast of Greenland at 66°N 36°W. The maps were carved from wood by the Tunumiit hunter Kuniit from Ammassalik. The short map shows the coastline, while the longer map represents islands along the coast.

researchers have been able to identify many of the places depicted. For example, Zaliv Lavrentiya, or St Lawrence Bay, on Chukchi Peninsula in present-day Russia, appears on the upper part of the Siberian Yupik map.

The Siberian Yupik map is not made to scale; nor does it have an 'up' or 'down', instead shifting viewpoints between different places. Because of this, some have chosen to call it a 'pictograph' or 'drawing' instead of a map. However, it is perhaps better to acknowledge that different cultures have different ways of representing places than to try to fit everything into one ideal of what a map ought to be.

A similarly innovative mode of mapping is seen in a set of nineteenth-century maps by the Tunumiit hunter Kuniit from Ammassalik, on the southeastern coast of Greenland (FIG. 49). Kuniit's maps are three-dimensional objects carved from wood. This means that they both portray an outline of the coast and mark out height differences of fjords and promontories. Kuniit sold the finished objects to the Danish explorer and naval officer Gustav Holm, who led an expedition to Greenland in 1883–85 that had the dual aim of making a geographical survey and determining the location of the medieval Norse settlements. Holm claimed that maps such as these were common, though it is hard to confirm this statement since Kuniit's maps are unique in being preserved into the modern period. Holm noted that the people he encountered 'knew their home region to the point' and that they would seek him out to look at the maps he had brought from Denmark, correcting both them and the maps Kuniit had made.[34]

The shorter of Kuniit's maps portrays a stretch of Greenland's coastline at 66°N 36°W. Let us start at the upper left-hand corner. Moving down the side of the map takes us north along the coast. Skirting the lower side, we continue further north while moving up on the right-hand side of the map. The second of Kuniit's maps depicts a set of islands located along the coast from the first map. Each island is connected by a narrower piece of wood so that they appear like a string of beads, though in reality the islands are not located on a straight line. Holm noted that Kuniit would move the two maps in relation to each other to show the actual position of the islands with regard to the coast.[35] In other words, to understand these maps fully, the user needed to possess a detailed mental map of the area as well as the ability to interpret the topography represented by the wood carvings.

At first glance, Kuniit's maps look like they show a bare coastline and thereby play to the idea of an empty north. However, the minute carvings on the map mark out abandoned human settlements that were used as storage sites. For example, one such site is visible on the south-western tip of the island Ananak or Depotø, the third island on the longer map. This means that the map portrays layers of human activity as well as a natural landscape. Brought back to Copenhagen, Kuniit's maps became part of the larger Danish colonial project of surveying Greenland and establishing relations with its inhabitants with the eventual aim of making the whole of Greenland Danish.

Maps made by indigenous people also provide evidence of how the use of land and sea changed with the weather and the time of year. One example of this is a map drawn in 1822 by the Inuk woman Iligliuk (FIG. 50). The map depicts a region south of Baffin Island, which is now part of northern Labrador in present-day Canada. The map shows as much of the coast as of the sea and through this landscape run dotted lines representing the trails the Inuit used to travel to get from one place to another. The lines on the map indicate sledge routes on the ice, with night-time camps on the shore marked by dots. Legends on the map provide information about where the sea was 'open at times' and where the Inuit would then use kayaks instead of sledges. In this way, the map, although it is not marking out cities or highways, provides a view of human life around Baffin Bay in the early nineteenth century. It is likely that the trails Iligliuk noted on her map had a longer history, having been used for many generations already when she drew it. Indeed, many of these trails are still in use today.[36]

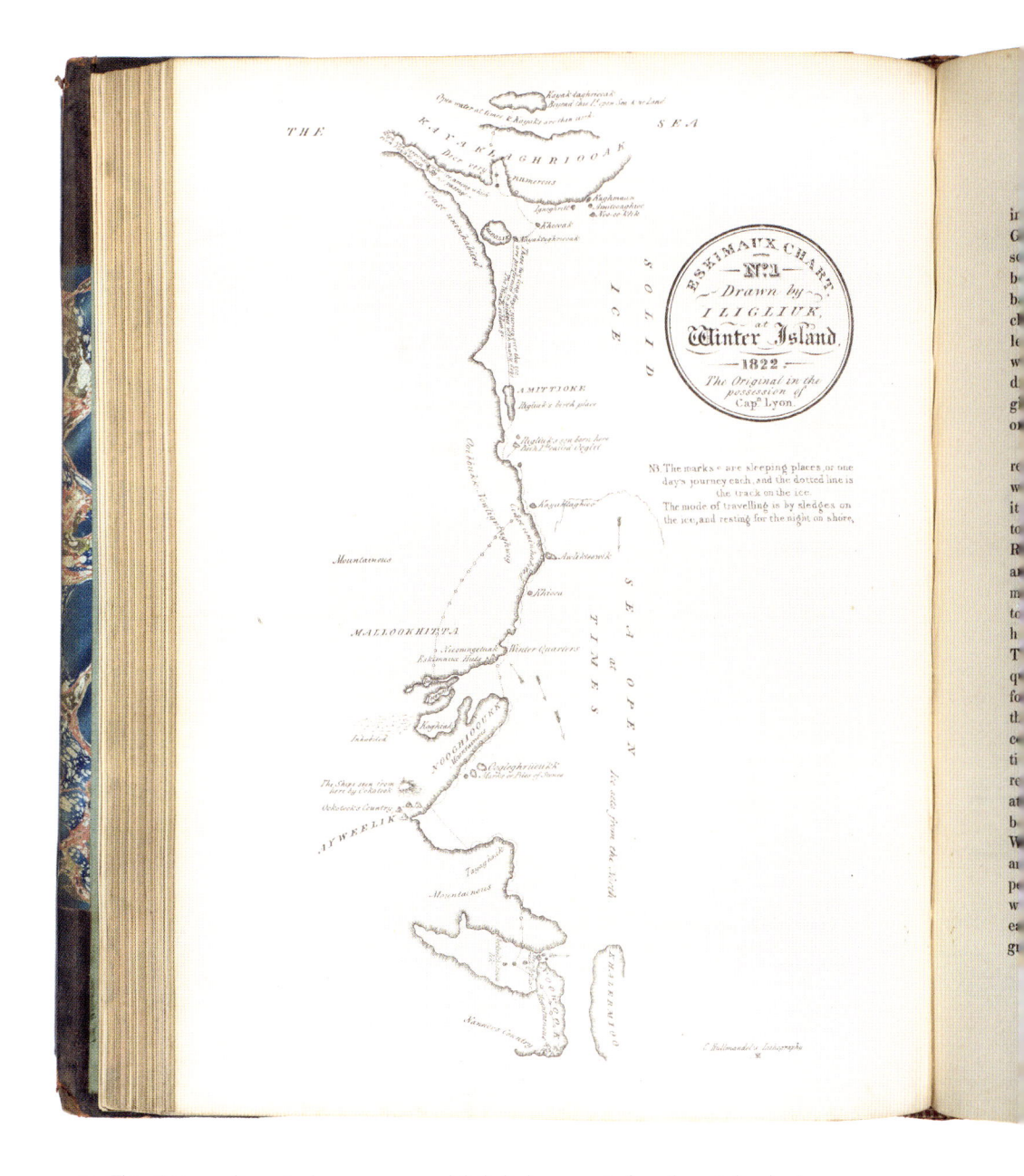

50 This 1822 map by an Inuk woman named Iligliuk shows part of northern Labrador in present-day Canada. Plotted along the coast are sleigh routes of winter passage and notes about where open water will require the use of kayaks.

Iligliuk, Kuniit and the anonymous map-maker from the Siberian Yupik give tantalizing indications of the creativity and range of possible representations at work when making maps. There is also evidence that indigenous maps were made using less durable materials, like drawing in snow or sand.[37] Whether these maps presented an empty or busy north is difficult to say, though if they were detailing human use of the landscape it is more likely to have been the latter than the former.

MAPPING THROUGH ENCOUNTERS

While the concerted efforts to survey the north from the early modern period onwards were directed primarily through the initiative of corporations and states bordering the Arctic to the south, expertise on northern geography was found locally in the north. When knowledge of geography grew among outsiders it was the result of peoples indigenous to the Arctic who aided – willingly or not – European, Russian and American explorers.[38]

For example, Iligliuk drew her map for the British explorer Sir William Parry on his second journey through Hudson Bay, which he was making in search of the fabled Northwest Passage from Europe to Asia. Iligliuk visited Parry's camp on Winter Island, in present-day Nunavut, Canada, in 1822. Parry noted in his journal that the Europeans drew an outline of the parts of the coast that they were familiar with and placed it in front of Iligliuk, who would then continue drawing. According to Parry, 'never were the tracings of a pencil watched with more eager solicitude'.[39] It was vital to Parry to find out whether he would be able to sail further west. Probably, that impetus made Parry and his fellow travellers ask for certain information and urge Iligliuk for more detailed descriptions regarding certain geographical features. Iligliuk would not have made this map, or at least not drawn it in this exact manner, had Parry not asked for it. In this sense, the map is the result of an encounter, and reflects both Parry's and Iligliuk's conceptions of the land, particular priorities and specific ideas about mapping conventions. Still, place names and trails similar to the ones Iligliuk noted on her map appear on other Inuit-made maps, and information about the area's land use can be corroborated through other sources. It seems that, even though mediated through Parry, the map provides some insights into indigenous conceptions of the north.

The steady stream of British explorers who attempted to find the Northwest Passage north of Canada to Asia in the nineteenth and early twentieth centuries – such as John Franklin, Robert McClure and Roald Amundsen – often portrayed the information exchange with people indigenous to the Arctic in amicable terms. A drawing by the British naval officer James C. Ross of a meeting between himself and the Netsilik man Ikmallik is no exception to this trend (FIG. 51). The

51 The British naval officer James C. Ross invited Ikmallik, of the Netsilik Inuit, on board his ship *Victory* to draw a map of parts of the Gulf of Boothia in northern Canada. The resulting map represents Ikmallik's knowledge of surrounding geography, but it was also shaped by Ross's expectations and wishes.

ENCOUNTERS & EXPLORATION   137

meeting took place in 1830 on Ross's ship *Victory* anchored on the west coast of the Gulf of Boothia. In Ross's rendition of the scene, Ikmallik is seen smiling while adding to a map begun by Ross.

However, the reality was most likely more complex. An indication of this is seen in the letters of the British explorer Sir Francis Leopold McClintock, who headed the Arctic expedition which, in 1859, finally established that the explorer Sir John Franklin had died in the Canadian Arctic more than a decade earlier. Franklin had set out on a mission to find the Northwest Passage to Asia in 1845, but the members of the expedition perished after the two ships, HMS *Terror* and HMS *Erebus*, got stuck in the winter ice. The loss of the Franklin expedition caught the imagination of the British public, and several new expeditions were sent out to ascertain what had happened.[40] McClintock was heading one of these expeditions; as a part of his search, he asked for information from the Inuit he encountered. To begin with, McClintock was interested in learning whether the Inuit knew anything about Franklin, but he also sought their knowledge on geography. Writing home to Franklin's widow, McClintock complained that 'It was a long time before I could get these Exquimaux to draw the cart.'[41] According to McClintock, the two Inuit chiefs who were assisting him, Nu-luk and A wa lak, did not wish to draw a map of a particular part of the coast that they considered unlucky, but McClintock persisted and eventually convinced them to do so.

It is difficult to know exactly what transpired here. McClintock likely wanted to appear competent to Lady Franklin, showing how he had overcome difficulties and how he had exerted himself to find information about her husband. At the same time, the episode also gives an indication of the agency of Inuit informers like Nu-luk and A wa lak.

52 Published in 1927, this British Admiralty map marks out the locations of finds from Sir John Franklin's ill-fated expedition. The map-maker has made a distinction between information from explorers (red) and from Inuit people (blue).

CHART SHOWING THE VICINITY OF
KING WILLIAM ISLAND
with the various positions in which relics of the Arctic Expedition under SIR JOHN FRANKLIN have been found.

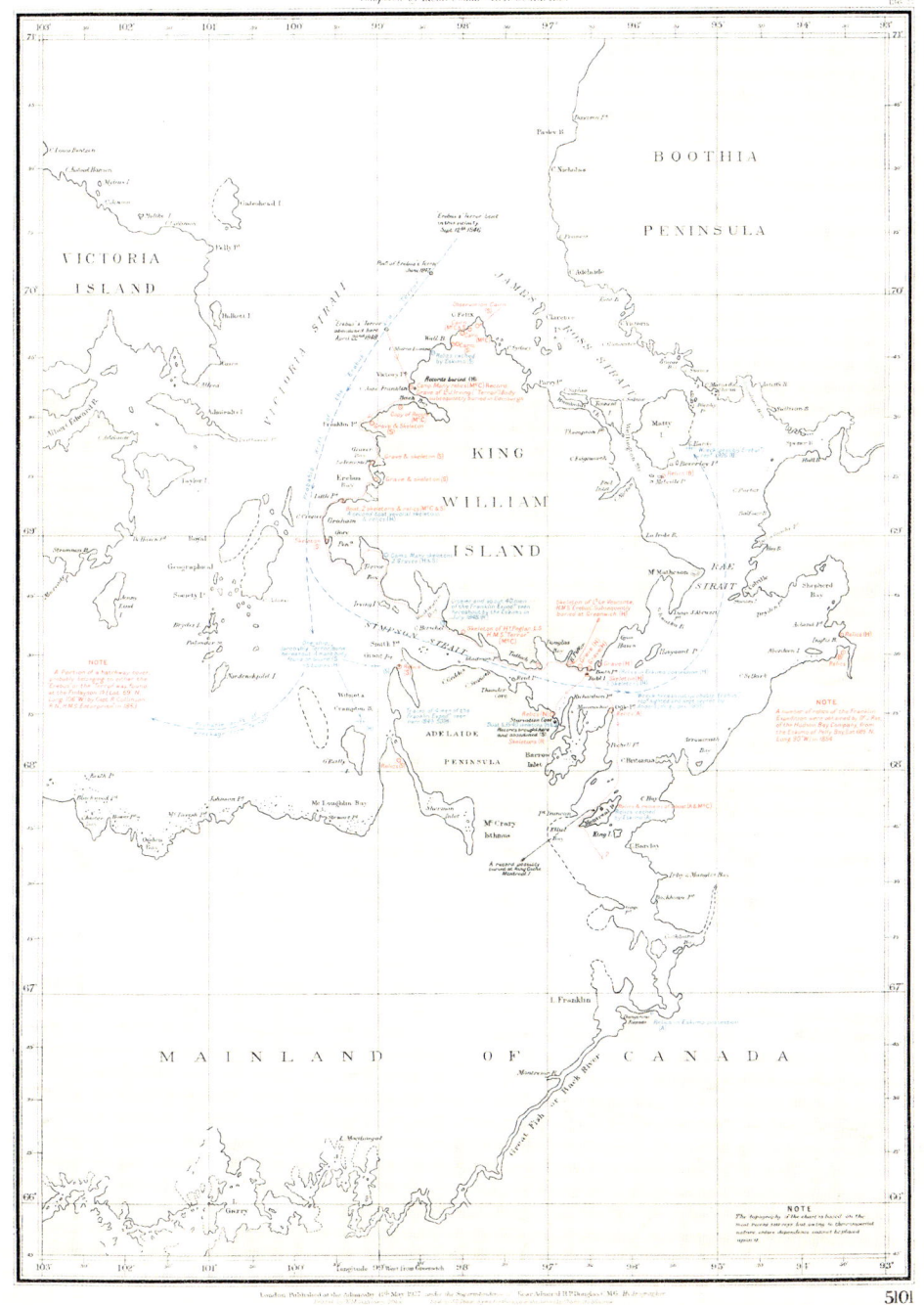

Compiled by Lieut. Commr R.T. Gould, R.N.

The two men approached McClintock, yet they were pressured into providing information they did not want to divulge. Moreover, they no longer had sole control of the information once they gave it to the British explorers.

Paradoxically, the expeditions sent out to find Franklin resulted in greater increases in knowledge about the north-west of the Canadian Arctic than what would likely have been the result if Franklin had been able to return home. The ambiguous role of Inuit knowledge in this process is clearly seen in a British Admiralty map from 1927, which indicates the places where remains from the Franklin expedition had been found (FIG. 52). The places reported by British and American explorers are marked out in red on the map. The locations marked out in blue signify finds by Inuit informers, and the map notes that this information 'probably is not altogether trustworthy'.[42] This comment might have been intended to signify that the information gained from the Inuit was reported second-hand in Britain; yet it is probable that it also reflects prejudice on the part of the map-maker. The statement seems particularly ill-advised given recent developments in the search for HMS *Terror* and HMS *Erebus*. The two ships were finally discovered south and south-west of King William Island in north Canada in 2014 (HMS *Erebus*) and 2016 (HMS *Terror*). That the ships could be located was a direct result of a revaluation of Inuit reports from the nineteenth century which had correctly identified where the ships had sunk.[43]

The word 'encounter' can entail anything from peaceful interactions to brute-force coercion. Most mapping encounters in the far north occurred somewhere between these extremes, but it is worth remembering that even in peaceful encounters power dimensions were at play and influenced the mapping process – not least because the control of the narrative to audiences further south lay almost solely in the hands of the southern explorers, companies and state interests. When we think of encounters as crucial to mapping processes, these should not be understood as interactions on equal terms.

MAPPING THE NOT SO EMPTY NORTH

Maps are made by people, and they are used by people. Consequently, maps give evidence of human activities. This is true even for a map of a supposedly 'empty' natural landscape, like some of the maps that depict the Arctic. In fact, very few parts of the world are untouched by human presence. Most landscapes include layers of human presence, as evidenced in Kuniit's map of eastern Greenland which showed abandoned settlements along the coast. And even when the north was conceived as empty, outlines of the Arctic coasts and islands were named after people, towns and states. Arguably, world and regional maps, by their very nature, promote views of an inhabited Earth. While there are many maps that have claimed as their subject matter the *geographical* outlines of the terraqueous world, almost all also give some information about the names that people have given to coasts, mountains, glaciers, islands and the like. Most maps also detail the borders of states and locations of major towns, and some give much more information about human life and priorities.

For maps of the north, two additional phenomena contribute to the image of a populated north. The first is the medieval and early modern conception of the north as a liminal place where wondrous beings might live. The manifestation of those ideas resulted in, quite literally, people on maps. The second way in which maps emphasize human presence in the north is the habit of detailing the progress of explorers' journeys on the maps of the far north from the early modern period onwards. However, it is also clear that while both features have contributed to maps that show human activity in the north, these maps have also contained within them exclusions of indigenous societies.

NEWFOUNDLAND AND NOVA SCOTIA

COD FISHERY OFF NEWFOUNDLAND

J. & F. TALLIS, LONDON & NEW YORK.

The Map Drawn & Engraved by J. Rapkin

ANIMALS ON NORTHERN MAPS

O LAUS MAGNUS filled both the land and the sea of his map of northern Europe with animals. He famously depicted whales and sea monsters splashing about in the North Sea and reindeer trotting across mainland Scandinavia. Other map-makers have perhaps been less exuberant in placing animals on maps, yet animals have never been far removed from the mapping of the north. Exploring the roles of animals in the mapping process is another way to probe the presumed idea of an empty north, while also highlighting how conceptions of the north have intersected with the exploitation of natural resources.[1]

We begin here by considering broadly the different ways in which animals were involved in the mapping of the north, and then move to a more in-depth discussion of three types of animal: codfish, whales and polar bears. There are interesting differences between these in terms of what they have meant for map-making and how they have been portrayed. While codfish featured as a commodity on the very earliest European maps of the North Atlantic, made in the late fifteenth and early sixteenth centuries, the whale first appeared as a sea monster and was transformed over time into a symbol of monetary value and

53 With changes in scientific and aesthetic ideals, illustrations moved towards the margins of many maps in the early modern period. However, the fish caught on lines and the two dogs resting among snow-covered branches still set the scene for this nineteenth-century map of Newfoundland and Nova Scotia by R.M. Martin.

54 Encounters between animals and humans in ritual and daily life decorate this bow drill made by an Inuit artist from the Bering Strait in the early nineteenth century.

scientific curiosity. And while polar bears both today and in a long tradition reaching back to the Middle Ages have been marked as exotic emblems of the wild north, the nineteenth-century expeditions to the Arctic also transformed the polar bear into a commodity, a companion and an object of study. Clearly, it is a misconception that map-makers placed animals on maps simply to fill gaps.[2]

THE ROLE OF ANIMALS IN THE MAPPING PROCESS

In the mapping of the north, animals appear in almost all aspects of map-making. To begin with, animals were, and are, an important theme in the art and symbolic mapping of indigenous societies in the far north. For example, the Siberian Yupik map of the Bering Strait, encountered in the previous chapter, shows elaborate scenes of whale hunting. This is not surprising. For much of history, people in the Arctic depended on fishing and hunting animals for their sustenance, for clothes, for tools and for trade. Animals were part of people's day-to-day life, and they were part of their spiritual life as well.[3] Mammals, fowl and fish appear as decorative elements in indigenous art. Scenes of hunting were, and are, common in bird's-eye views that adorn valued objects, like carved bow drills. The drill depicted here (FIG. 54) was made and decorated with scenes from daily life and spiritually significant events by an Inuit artist from the Bering Strait region in the early nineteenth century.[4] Caribou, fish, dogs and birds are portrayed next to people hunting, fishing and cooking. Scenes such as these are not strictly speaking representations of actual places, but rather a kind of spatial conception of the world and how humans and animals fit into it.

Animals were similarly common on medieval and early modern maps from Europe. However, while indigenous depictions portrayed animals that were part of the map-makers' everyday lives and religious world-views, the animals on medieval and early modern European maps were more commonly symbols of the exotic and represented natural and zoological riches. Often, these representations straddled the line between actual animals and creatures of the imagination. As with the depictions of people, fewer animals made it onto the actual maps when stylistic preferences changed in the seventeenth and eighteenth centuries. Still, animals lingered on the margins of maps as decorative elements that emphasized the character of the depicted land. Zoogeographer Wilma George notes that it was often the large mammals, birds and reptiles that were put on maps to represent a part of the world, and that their role seem to have been to emphasize difference rather than similarity between the familiar and the new and distant lands.[5]

That said, the depictions of certain species of animals on maps just as often mirror underlying economic priorities or conceptions of legitimate scientific inquiry. Such factors not only incentivized map-makers to include polar bears, codfish or whales as decorations on the margins of their maps, but also dictated what information map-makers included on the map itself and what was left out. Indeed, the possibilities of

ANIMALS ON NORTHERN MAPS 145

economic gain often determined what regions should be mapped in the first place. In this process, the map medium, at times erroneously, gave animal species geographic coordinates and presented the north as a region of inexhaustible bounty.

Another way in which animals were central to the mapping of the north was in their use as raw materials for the maps themselves. Before the gradual introduction of paper in the Middle Ages, European maps and texts were commonly drawn and written on parchment made from animal skins. For example, figure 55 shows an early-fourteenth-century manuscript made of parchment, depicting the thirteenth-century English monk Bartholomaeus Anglicus writing on another parchment. Maps for luxury consumption continued to be made from animal skins well into the eighteenth century. In East Asia and the Arab world, paper gained popularity earlier. In the Arctic, where resources are scarce, animal products continued to be used for decorative arts longer, as seen in the nineteenth-century bow drill.[6]

An added dimension to the role of animals in the Arctic expeditions of the nineteenth and early twentieth centuries is that animal labour and produce were essential for the exploration of the inhospitable frozen landscape of the far north. When Robert E. Peary contemplated what had been most important for his 1909 expedition to the North Pole, he included on his list the dogs that had drawn the sledges, and

55 Parchment made of animal skins was used in medieval Europe for both texts and maps. This historiated initial on an early-fourteenth-century manuscript, made of parchment, depicts the monk Bartholomaeus Anglicus writing on another parchment.

the clothes and gear made of animal skins that Inuit women had sewn for the crew. Peary claimed that

> Man and the Eskimo dog are the only two machines capable of such adjustment as to meet the wide demands and contingencies of Arctic travel. Airships, motor cars, trained polar bears, etc., are all premature, except as a means of attracting public attention.[7]

Peary relied on local knowledge when deciding to use local sledge dogs, and when adopting the dress of the Inuit.[8] Incidentally, he does not mention local knowledge of *geography* in his list of success factors, though we have seen that the Inuit also played an important role in knowledge transfer regarding geographical features. Peary's recognition of Inuit expertise instead focused on how they utilized the scarce resources of the land to survive in a harsh climate. When he was

56 Being appropriately dressed is essential for Arctic travel, as the American explorer Robert E. Peary well knew. Inuit women sewed clothes from local animals for him and his crew. Plate from *The North Pole* by Robert Peary, 1910.

describing his polar exploration, the theme of a land of bounty was clearly not relevant. This highlights the diversity of northern natural environments, and by extension the diversity of conceptions about the north. When it comes to the mapping of the north, animals were at times highly visible as subject matter or decoration, yet, even when no animals appear on a map, there were likely human–animal relationships behind the scenes that influenced the mapping process.

CODFISH AND THE MAPPING OF THE NEW WORLD

At first sight, the text on a Spanish 1529 world map (FIG. 57) seems dismissive. The map was made by the cosmographer Diego Ribero (or Diogo Ribiero in his native Portugal). A note on the west coast of present-day Canada reads 'Tiera Nova de Cortereal in which there is no other benefit than fishing for codfish and lots of pine wood'.[9] Read out

57 The Portuguese map-maker Diego Ribero worked for King Charles V of Spain. His maps showcase Iberian geographical knowledge about the newly encountered land and sea in the North Atlantic. This detail from *Carta Universal* (1529) shows a stretch of the coastline of present-day Canada and includes references to the increasingly important codfish.

MAPPING THE NORTH

of context, this sounds like the lament of a disappointed tourist, expecting to find exciting architecture and pleasant beaches, and instead encountering a backwater where fishermen go about their business and forest covers the shores. In reality, the comment is a reference to an industry of high value in the early sixteenth century, and one of the driving forces behind the European exploration and mapping of North America: the cod fisheries off the eastern coast of North America, centring on Newfoundland in present-day Canada. As noted by K.J. Rankin and Poul Holm, it is not a coincidence that so many of the early maps of the eastern coast of North America mention fishing, even including outlines of the most important fisheries.[10] Tellingly, the Portuguese referred to this part of the world as the Land of Bacalhao, a name referencing the Portuguese word for dried or salted codfish, which was, and is, popular in Mediterranean cuisine.[11]

In the standard narrative the explorations of Europeans westward in the early modern period are explained as attempts to find new sea passages to the lucrative spice markets of East Asia, or as driven by a wish to explore and exploit the riches of the interior of the new continent. An equally important, though less recognized, factor of westward exploration in the sixteenth century was the lure of the newly discovered fishing grounds along the north-eastern coast of North America. From around the year 1500, fishing boats from a number of regions in western Europe began to find their way to the shallow waters surrounding the islands off the eastern coast of present-day Canada in ever larger numbers. In particular, Iberian fishermen were quick to make use of these new fishing grounds. The fresh fish was salted or dried on site, and then prepared to bring back to a European market hungry for cod. In what has been called the 'Fish Revolution', the catch volumes of cod from the North Atlantic saw an estimated fifteenfold increase in the decades following 1500.[12]

The economic importance of cod as a commodity led to images of fish and fishing grounds appearing on European maps. For example,

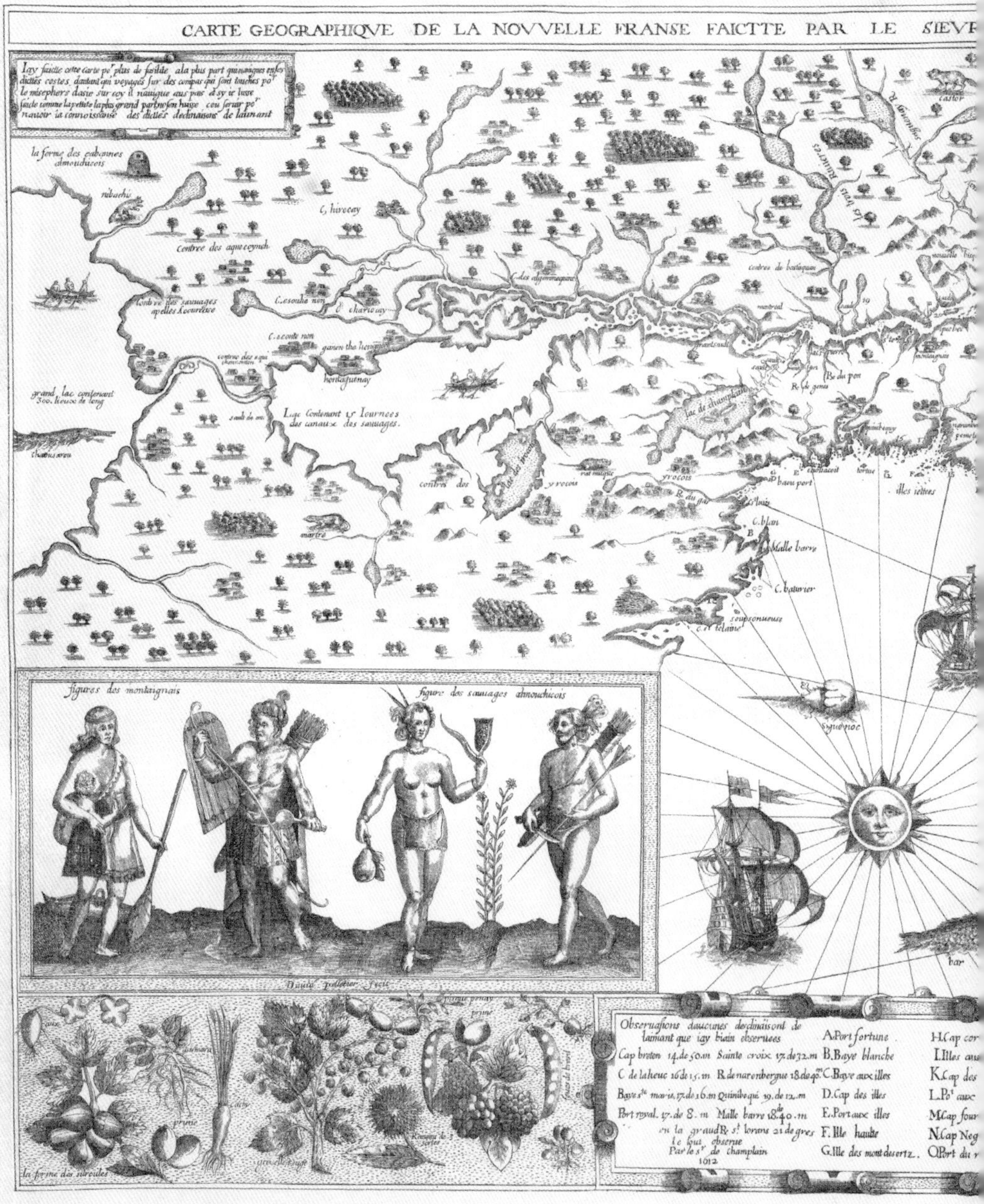
CARTE GEOGRAPHIQVE DE LA NOVVELLE FRANSE FAICTTE PAR LE SIEVR
Iay faictte cette carte po.r plus de facilite ala plus part qui naviguent en les dictes costes, dautant qui voyages sur des compas qui sont touches po.r le misephers dasie sur soy il navigue aus pais et sy ie lusse faicte comme la petitte la plus grand part ne sen husse cou servir po.r navoir la connoissanse des dictes declinaisons de laimant
la forme des cabannes almouchicois
ribachis
Contree des aquecoynch
Contree des sauuages apelles Kourekese
C. esouhe non
C. charioay
C. hireceay
C. des algommequins
Contree de bentaquan
montreal
lac de champlain
grand lac contenant 300. lieux de long
Lac Contenant 15 Iournees des canaux des sauuages
saulx de ons
martre
contre des
yrocois
R. du gas
C. blan
Malle barre
C. baturier
souspsonueuse
C. st. telaine
Syguenoc
castor
figures des montaignais
figure des sauuages almouchicois
Dauid pelletier fecit
la forme des sitroulles
Obseruasions daucunes declinaisont de laimant que iay biain obseruees
Cap breton 14. de 50. m Sainte croix 17. de 32. m
C de laheue 16 de 15. m R. de narenbergue 18 de 40. m
Bayes.te maris. 17. de 16. m Quinibequi 19. de 12. m
Port royal. 17. de 8. m Malle barre 18 de 40. m
en la grand R. st. lorans 21 de gres le tout obserue Par le s.r de champlain
1612
A. Port fortune
B. Baye blanche
C. Baye aux illes
D. Cap des illes
E. Port aux illes
F. Ille haulte
G. Ille des mont desertz

MPLAIN SAINT TONGOIS CAPPITAINE ORDINAIRE POVR LE ROY EN LA MARINE
faict len 1612
labrador
Canadas
lauteur n'a point encorre recongnu cette coste
Golfe st lorans
la grand baye
baye sauvage
Terre Neuue Ille
Mons de Nostre dame
Canadiens
Ille aux taugeux
baye st iulian
ille de bacalles
baye de la conseption
ban neuf
Ille de sable
banqureaux
lou marin
le Grand ban
saumon
ballaine
chien de mer
moluë
gros chabot
sturgon
lieux
1612
Auec priuilege du Roy
Y. Port de saualette
Z. Pasage du glas
&. Port aux anglois
2. Baye courante
3. Cap de percé
4. Ille grande
5. Pasage couran
6. Baye de gennes
7. Ille perdue
8. Cap de mine
9. Port aux coquilles
10. Illes iumelles
11. C. st Iean
12. Langille
13. Laheronniere ille
14. Illes rangees
15. Baye st luct
16. Passage du gas
17. Coste de mommorancy
18. R de champlain
19. R sainte marie
20. Ille dorleans
21. Ille de baqus

Samuel de Champlain's early-seventeenth-century map of the French possessions in North America (FIG. 58) displays dotted outlines of the major fishing banks and engravings of various kinds of fish.[13] As a driving force for exploration, the codfish in the Atlantic Ocean was also important for the mapping of North America in that information from fishing expeditions contributed significantly to knowledge about the geography of the north-western Atlantic among map-makers in southern Europe. North European countries like Norway and Iceland had long had a stake in the North Atlantic fishing markets. The 'discoveries' of new lands and fishing grounds in the north-west gave an opportunity for new actors to move in. This meant that knowledge about this part of the world grew, but also that the new knowledge was steeped in the interests and priorities of the fishing expeditions and their European financial backers. Stephanie Pettigrew and Elizabeth Mancke point out that the potential economic profits from cod fishing in the Atlantic Ocean made Iberian map-makers like Diego Ribero pay attention to the north-western Atlantic islands, while focusing considerably less on getting the outline of Scandinavia right.[14]

The competition to control North Atlantic fishing was intense. Samuel de Champlain's map highlights this by showing not just fishing grounds but also silhouettes of large galleys firing their cannons. In the early seventeenth century, when Champlain made his map, the French were investing in North America, establishing religious missions, colonies, fisheries and trading posts.[15] Through his numerous expeditions and the publication of travel accounts and maps, Champlain was a central figure behind many of these entrepreneurial endeavours. He was also influential in projecting the New World as a bountiful land waiting to be explored.[16] It is worth noting that the impression given

58 PREVIOUS SPREAD The French map-maker Samuel de Champlain has marked out the fishing grounds off the coast of present-day Canada with dotted lines and shading. The map further emphasizes the bounty of the New World through the illustrations of fish in the sea and botanical specimen at the bottom margin.

by Champlain's map of New France is not a land of cold and physical hardships, but instead a land of rich natural resources. The engravings of half-clad indigenous people evoked a temperate exotic climate, rather than a cold north. In creating interest in France for the North American venture, this mode of representation was likely more effective than a portrayal of the harsh winters Champlain had himself experienced. Like Olaus Magnus and Abraham Ortelius, Champlain chose to portray one season over another, thereby foregrounding one kind of climate in his depiction of the newly 'found' land.

Moving forward in time, the eighteenth-century British map engraver and publisher Thomas Jefferys highlighted on his maps of North America – just as Ribero's and Champlain's maps had – the fisheries off the coast of Newfoundland. Figure 59 shows a compilation map based on Jefferys' cartographical work but revamped by his heirs and published in its first edition in 1776 under the title *An Accurate Map of North America*.[17] The dotted outline of the Newfoundland Grand Banks at the top of the map makes clear that fishing was still an important industry when this map was made. However, the power dynamics had changed dramatically, as had the European presence in, and knowledge of, North America. In the second half of the eighteenth century the competition between European powers for domination in North America was an acute issue, and maps were a part of this. Map historian J.B. Harley notes that, through his maps, Jefferys 'became a geographical mouthpiece' for British imperial strategies regarding America, promoting British claims over French and Spanish interests.

One way in which *An Accurate Map of North America* was framed around British perspectives on North America was through the inclusion of text extracts from the 1763 Peace Treaty of Paris, which ended the Seven Years War. This was a global conflict fought between Britain and the combined forces of France and Spain. The rights to the Newfoundland fisheries were a central issue in the peace negotiations. As can be gleaned from the text on the map, Spain conceded all rights

A Particular Map of BAFFIN and HUDSON'S BAY
THE ARCTIC CONTINENT
BAFFIN'S or BYLOT'S BAY
HUDSON'S BAY
NEW NORTH WALES
NEW SOUTH WALES
NEW BRITAIN
LABRADOR
NORTH SEA
HUDSON'S STRAITS
NEW SOUTH WALES
JAMES BAY
ALGONQUINS or ADIRONDAKS
LAKE HURON
LAKE MICHIGAN or ILINOIS
L. ONTARIO or CADARACUI
LAKE ERIE or OSWEGO
PENSILVANIA
MAHA
TINTONS
SIOUX
WHITE PADUCAS NATION
LOUISIANA
PANIS NATION
KANSEZ NATION
BLACK PADUCAS NATION
OZAGES NATION
VAST TRACT OF LAND UNKNOWN
ILINOIS
CHIKTAGHEEKS
VIRGINIA
NORTH CAROLINA
SOUTH CAROLINA
GEORGIA
TEGUAYO GRANDE or TEGUAS NATION
NEW MEXICO
APACHERIA
APACHES NATION
PIMERIA
NEW NAVARRE
CALIFORNIAN GULF
NEW KINGDOM of LEON
GULF OF MEXICO
BAHAMA or LUCAYOS ISLANDS
HISPANIOLA
GREAT ANTILLES
CARIBBEAN
HONDURAS
CHIAPA
NICARAGUA
COSTA RICA
THE PACIFIC OCEAN OR GREAT SOUTH SEA
CALIFORNIA
NEW MEXICO
SEA OF CALIFORNIA
PART OF CALIFORNIA
NEW NAVARRE
SOUTH SEA

to them at the end of the Seven Years War.[18] France also made significant concessions, giving up all of its land possessions in continental America east of the Mississippi river, except New Orleans, which France held on to a while longer. However, the French were allowed to continue their fishing activities from the Grand Banks, and the two small islands Saint Pierre and Miquelon were ceded to France so that their fishermen could dry their catch. These islands are still French dominions. However, due in part to the collapse of the codfish stock in the area in the 1990s, the local fishing industry has been much reduced.

Fishing was as central an industry in Norway as it was in Newfoundland across the North Atlantic Ocean. A poem written in 1739 by the Norwegian clergyman Petter Dass prophetically highlights

59 Published in its first edition on the eve of the American Declaration of Independence of 1776, this map by Thomas Jeffreys shows North America as a British domain. Later editions were hastily changed to distinguish between 'The United States, & the Several Provinces and Colonies which Compose the British Empire'.

ANIMALS ON NORTHERN MAPS 155

a nation's vulnerability when relying on an exhaustible resource. Dass asks, 'Should the codfish fail us, what had we then? / What would we bring to [the market] in Bergen? / Then the ships would sail empty.' Concerns for the health of the codfish led to several Norwegian mapping projects in the nineteenth and twentieth centuries. These were maps and sea charts of increasing accuracy. Norway took a leading role in the making of sea charts in the late nineteenth century. Underlying these endeavours were, among other factors, economic incentives to facilitate fishing.[19]

THE WHALE FROM MONSTER TO COMMODITY

The North Sea is home to sea creatures larger than ships and fiercer than fleets, if we are to believe Olaus Magnus. His depictions of animals in the sea include both large whales and creatures of the imagination. For instance, on the middle left side of *Carta Marina* a stately creature lifts its head from the waves and sends two spurts of water high into the air (FIG. 60). Olaus Magnus described this animal both as a 'monstrous fish' and as belonging to the family of whales, and that it was called *Physeter* or *Pristis*. He detailed how the animal preyed on ships, showering them with water or squeezing them with its long tail, concluding that it 'exposes seamen to the severest danger'.[20]

In fact, no actual whale resembles Olaus Magnus's *Physeter*, half raised out of the water and with multiple fins on its back and a tongue sticking out of its mouth. Still, Olaus Magnus's sea creatures were not simply fictious embellishments on his maps. He had travelled through the Nordic region, he recounted how others had seen stranded whales and encountered them at sea, and he could report on the details of the whaling industry.[21] Through this ambiguity between monster

60 Olaus Magnus depicted monstrous sea creatures in the North Sea. He writes about the large whale that lifts its head above the waves: 'Among the family of sea-monsters the spotter, or leviathan, three hundred feet long, has been given a nature that is totally forbidding.' Detail from *Carta Marina*, 1572.

Hamburgen.

M

Scoti

Vespemo

Ostrabord

Bremen

Lubicenses

O

Pistrum siue Phiset

K

Monstrum 1537 visum.

FARE

Hordero

Sudero

B

Eclea

Mulse

A

C Moachus

Stremb

TILE

Hec Insula habet 30 millia popul. et amplius Hec habitat Dñus Insularum

L

BALENA

ORCA

D

ORCADES XXXIII

M

Pomona

Epatus Org Regum Antiqua sepulture

Olim Regnum

H

N

I

Ski

AMERICÆ SEPTENTRIONALIA INCOGNITA

Briggs his Bay
Nova Dacia
Cape Philip
Port Nelson
Button's Bay
New Seaven
New North Wales
Baffins Bay
Sir Thomas Smiths Bay
Whale Sound
Sir Dudley Diggs Cape
The Iland
New South Wales
C. Southampton
Spirietta Maria
Cap. Pembroke
James his Bay
Hudsons Bay
Freetum Hudson
C. Charles
Cumberland Ile
Cheredbay
Nova Britannia
Canada
Nova Britannia

Freetum Davis
Lord Davis ile
Chidley cape
Meta Incognita
Desolatio
Canada
Terra Noua

DEUCA

TARTARIÆ
MARITIMA
INCOGNITA

TARTA
RIÆ
PARS

Lucomorye
BAIDA
Iougoreia
NAGAIA HORDA
KOLMACKI KOLMACKI
SIBIRIA
BULGARIA
NIAREN MORE
Nova Zemla
OBDORA
Astracan
CASAN
MOURMANSKOY MORE
Czeremissi Lugovoy
Czeremissi Nagornay
Pole
Spitzberge
DVINA
MORDUA
WAGA
Wologda
MOSCOVIA Rezania
OCEANUS
Dikoia
LAPPONIA
SUECIA
Finlandia
SEPTEN-
Severskigris
TRIONALIS
Ducatus
LITVANIA
Samogitia
MARK GERMANI- CUM
POLONIÆ PARS
GERMANIÆ PARS
SCOTIA
HI-BER-NIA ANGLIA

and animal, the whales on *Carta Marina* were renderings of real-life animals and at once figuratively symbolized the dangers of the open sea. Because relatively little was known about whales in mid-sixteenth-century Europe, Olaus Magnus's whale depictions gained considerable attention.[22]

One reason why whales seemed so fantastical is probably the difficulty involved in studying them in their natural habitat in the past. For much of history, people could either observe a whale from what was visible from the surface, or see a whole animal stranded on a shore. In the latter case, the whale carcass quickly begins deteriorating.[23] As a result, many early depictions of whales misrepresented the animals. There was also considerable variation in how the different species were named. While the *Physeter* was a near monstrous creature to Olaus Magnus, it was a *Nord Kaper* to the German map-maker and engraver Johann Baptist Homann.[24] Today this species is known as the sperm whale.

Eventually, people began to classify whales more as animals than as monsters.[25] Not least, scenes of whaling began decorating maps of the Arctic as the commercial gains from the whaling industry grew from the seventeenth century onwards. And though whales traversed both the southern and northern parts of the world, Homann's depiction of a *Physeter* whale appeared in the section of his *Grosser Atlas* on the Arctic. Similarly, the Dutch map-maker and publisher Henricus Hondius surrounded his map of the Arctic from 1636 (FIG. 61) with scenes of whale hunting and resource extraction.[26] The result is an impression that whaling was integral to the north. It is also worth noting that the whales on Hondius's map are much less detailed than the sea creatures on Olaus Magnus's map. To Hondius it was not the individuality of

61 PREVIOUS SPREAD Scenes of whaling surround Henricus Hondius's map of the Arctic from 1636, effectively connecting whaling to the north. A colourist has used different colours to mark out the coastlines and state boundaries. As a result, 'Lapponia' appears as an independent domain in northern Scandinavia.

the various species that was important, but rather the implications of economic potential that the scenes depicting whaling would evoke among contemporary audiences.

The whaling industry began to expand rapidly in the seventeenth century. John R. Bockstoce has described this development as the 'most powerful agent of change' in parts of the Arctic, transforming both the natural world and human societies.[27] The sought-after commodities of the whaling industry were oil and baleen. The oil was used for illumination and for soap; baleen is a flexible material that had a variety of everyday uses, such as in corset stays, umbrella frames and riding whips. The demand for commodities such as these in eighteenth- and nineteenth-century Europe and America were seemingly inexhaustible, but the whales were not. Elizabeth Ingalls describes this as a 'recurring historical theme, year to year and ocean to ocean: an initial abundance of whales, followed by overfishing, a decline and the search for new grounds, at which point the pattern begins again'.[28] From the Yupik map of the Bering Strait (see FIG. 48) we get a glimpse of the frenzy among whalers in the mid-nineteenth century to catch right whales and bowhead whales that migrated through the waters north of the Bering Strait.

Maps and prints like Hondius's map of the Arctic draw attention to the commercial importance of whale hunting while also giving the viewer a strangely abstracted and exotic view of the animals outside of their natural habitat. The images as well as the bird's-eye views were copied from earlier works, as was so often the case in early modern printmaking, and it was not unusual for errors to be introduced – not least because most engravers, though striving for pictorial accuracy, had never seen a whale in real life.

Scientific interest in physical geography and in documenting the lifespan of northern animals played an increasingly important role in expeditions to the north in the nineteenth century, though these scientific interests overlapped with commercial interests. Maps that

62　The Scottish map-maker Alexander Keith Johnston published an English edition of the German map-maker Heinrich Berghaus's monumental *Physikalischer Atlas* in 1849. The atlas details the geographical distribution of natural phenomena, including, as seen here, fur animals and whaling (on the insert map).

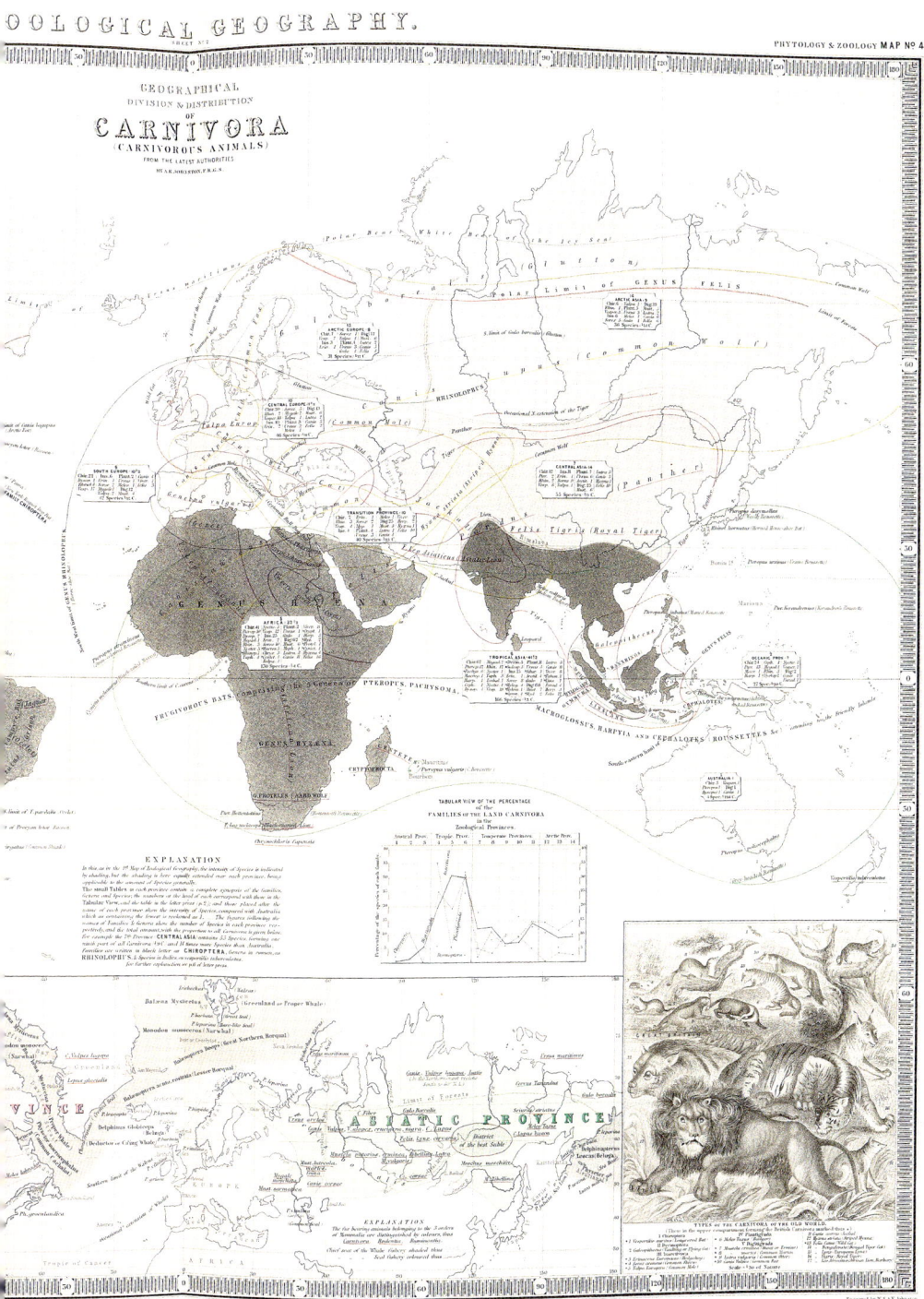

ZOOLOGICAL GEOGRAPHY.

PHYTOLOGY & ZOOLOGY MAP No 4

GEOGRAPHICAL DIVISION & DISTRIBUTION OF CARNIVORA
(CARNIVOROUS ANIMALS)
FROM THE LATEST AUTHORITIES
BY A.K. JOHNSTON, F.R.G.S.

TABULAR VIEW OF THE PERCENTAGE of the FAMILIES OF THE LAND CARNIVORA in the Zoological Provinces.

EXPLANATION

TYPES OF THE CARNIVORA OF THE OLD WORLD

Engraved by W. & A.K. Johnston

William Blackwood & Sons, Edinburgh & London.

combined information about animal habitats and fishing, such as the zoological maps made by the German cartographer Heinrich Berghaus, satisfied both interests.[29] For example, his map of the distribution of species of carnivorous animals across the globe includes a small inset map of the northern hemisphere containing information about the habitat of 'fur-bearing animals' together with the 'Theatre of the Whale and Seal Fishery' (FIG. 62). Berghaus's *Physikalischer Atlas* was a pioneering work in physical geography, which drew on Alexander von Humboldt's vision of scientific inquiry. One of Berghaus's aims was to give spatial coordinates to specific species of animals. However, once again, the whale eluded specific coordinates since its movements in the oceans were hard to determine. Instead, Berghaus focused his map on the whale and seal as commodities. Consequently, what is mapped is the relationship between the animal and its human hunters.[30]

HERE BE WHITE BEARS

The word 'Arctic' comes from the Greek word *arktikos*, meaning 'the land of the bear'. It is a reference to the constellation of stars called the Great Bear, seen in the north sky at night. While in the present day the polar bear has become a symbol in popular culture for the distress caused by climate change, it has a much longer history as an emblem of the dangers and the wilderness of the far north.[31]

The name 'white bear' was long used for the bear that is commonly known as the polar bear, *Ursus maritimus* today. References to 'white bears' appear in numerous medieval accounts on the subject of geography.[32] For example, the thirteenth-century scholar Bartholomaeus Anglicus – seen working on his influential encyclopaedia *De proprietatibus rerum* ('On the Properties of Things') in figure 55 – wrote that there were 'white bears most great and right fierce' in Iceland, and that they made holes in the ice to catch fish.[33]

The Latin phrase *Hic sunt ursi albi* ('Here are white bears') on the map-maker Angelino Dulceti's two nautical charts from the early fourteenth century announces some of the earliest appearances of polar bears on the face of a map. Dulceti, based in Majorca and Genoa, located the polar bears on the Scandinavian peninsula.[34] On the German map-maker Martin Waldseemüller's world maps from 1507 and 1516 the note about white bears has moved further away – to northern Russia.[35] Such references to the fauna of the far north can be found in several medieval and Renaissance manuscripts and on maps made in Europe and the Middle East, showing how geographical knowledge moved between places and societies.[36] The existence of strange and potentially dangerous animals was interesting information to include, and, like the information about whales, notes about polar bears were used to emphasize the exotic character of a region. In other faraway parts of the world a map-maker might write 'here be dragons', but in the north the curious and monstrous took the shape of a white bear.

As European states began to send expeditions to the far north in the sixteenth century, and as accounts of their adventures began to appear on the shelves of European bookshops, the white bear became less of a myth and more of a known feature of the northern regions. On early modern maps of the north of Europe, like Abraham Ortelius's *Islandia* and Olaus Magnus's *Carta Marina*, polar bears or *ursi albi* (Latin, 'white bears') are seen balancing on ice floes and gnawing on large fish. While it is true that polar bears live, when they can, mostly on the ice, their preferred dinner is seal rather than fish. And, despite notes on the maps indicating that this animal was white, the polar bears on many early modern maps were coloured distinctly brown by artisans after printing. Nor were map-makers sure what to make of the reports of white bears in newly encountered North America, at times colouring them blue or brown.[37] This animal was not well known to most of Europe in the early modern period.

This did not, however, diminish European fascination with the animal. On the contrary, readers were captivated by numerous accounts of near-death encounters with ferocious polar bears. The Dutch

Caerte van Nova Zembla, de Weygats, de custe van Tartarien en Rus-
landt tot Kilduyn toe, met anwysinge
van de weder vaert lancx de Noort-
cust van Nova Zembla, en de over-
vaert omtrent de Weygats na Rus-
landt, tot de hoeck van Candenos, en
de mont van de Witte Zee.

Door Gerrit de Veer
beschreven.
Baptista à Doetechum sculp. a° 1598.

Den Swarten hoeck
D'Admiraliteyts Eylandt
C. Plantio
Loms bay
Groote bay
Lange nes
D'eerste hoeck
Cants hoeck
Swarte klip
Costint sarch
Cruys hoeck
Schans hoeck
S. Laurens bay
Meel haven
Laech Eylandt
Het laeghe landt
Twe Eylanden

NOVA

Tretum Weygats

Kilduyn
Cola
Olena
7. Eylanden
LAP
Warsina
PIÆ PARS

Candenos
Colgoy
Toxar
RUSS
Pitsora
IAE PAR

Barentsz expeditions sent in search of the Northeast Passage in the last decade of the sixteenth century were one example of this. Gerrit de Veer, who took part in two out of three of the Dutch expeditions, described more than twenty encounters with polar bears during a nine-month period, including several that ended with the death of either man or bear, or both.[38] Three polar bears run across the tundra in northern Russia on De Veer's map of Novaya Zemlya (FIG. 63). The polar bears appear as relatively stylized outlines, but the copperplate prints that accompanied De Veer's account give a more vivid depiction of the animals. Close to half of the prints in the book portray polar bears, including

63 The Dutch Barentsz expeditions searched for a Northeast Passage to China in the late sixteenth century. This map from Gerrit de Veer's account of the expedition, published in 1598, shows their route along the coast of Novaya Zemlya and the house where they spent the winter. Three polar bears run across the Siberian tundra at the bottom of the map.

ANIMALS ON NORTHERN MAPS 167

Pourtraict du murtre miserable, fait par vn cruel, horrible, & devorant Ours, qui miserablement a deschiré deux des nostres: & comme par deux fois l'avons combatu a toute force, & estions long temps empeschez, pour le tuer: lequel si long temps qu'il estoit vivant, non obstant, qu'il estoit blessé, harquebusé & batu, ne voulut quitter sa proye.

one of a bear attacking and mangling a crew member, and another of the failed attempt to put a leash on a polar bear and bring it back on the ship. De Veer described a polar bear attack in vivid terms: 'The beare at the first faling vpon the man, bit his head in sunder, and suckt out his blood, wherewith the rest of the men that were on land, being about 20 in number, ran presently thither, either to saue the man, or else to driue the beare from the dead body.'[39] In De Veer's account, the bear clearly was the victor in the confrontation. These accounts likely helped shape the public imagination of the polar bear as aggressive, and, by extension, of the north as a dangerous place.

Over time, references to hostile polar bears became a fixture of travel narratives. Eventually the balance shifted, and people became a threat to the bears, rather than the other way round. The meat of a polar bear is edible. Given the opportunity, the Arctic expeditions of the nineteenth century shot enough bears and walruses to sustain them through the winter, and sometimes more. For example, the members of the British Jackson–Harmsworth expedition killed sixty polar bears during 1885 while exploring Franz Josef Land, an archipelago north of Siberia.[40] The leader of the expedition, Fredrick Jackson, mused that the hunt also provided a pleasurable pastime. After describing how a bear had nearly killed him, Jackson added that 'Altogether they [the polar bears] have afforded great entertainment during the winter, and have certainly done a great deal to relieve the monotony.'[41]

64 Gerrit de Veer's account of William Barentsz's expeditions to northern Russia and Novaya Zemlya (1598) includes numerous descriptions of encounters with polar bears. Often the bear got the upper hand, as seen here where 'a cruel, horrible, and devouring bear' killed two men.

65 William Barentsz and his crew captured polar bears and tried to leash them and bring them back to the ships.

66 OVERLEAF The British Jackson–Harmsworth expedition explored Franz Josef Land, north of Siberia. The expedition members killed sixty polar bears during the winter of 1885, as a pastime, in self-defence and as sustenance. Royal Geographical Society, 1898.

ANIMALS ON NORTHERN MAPS

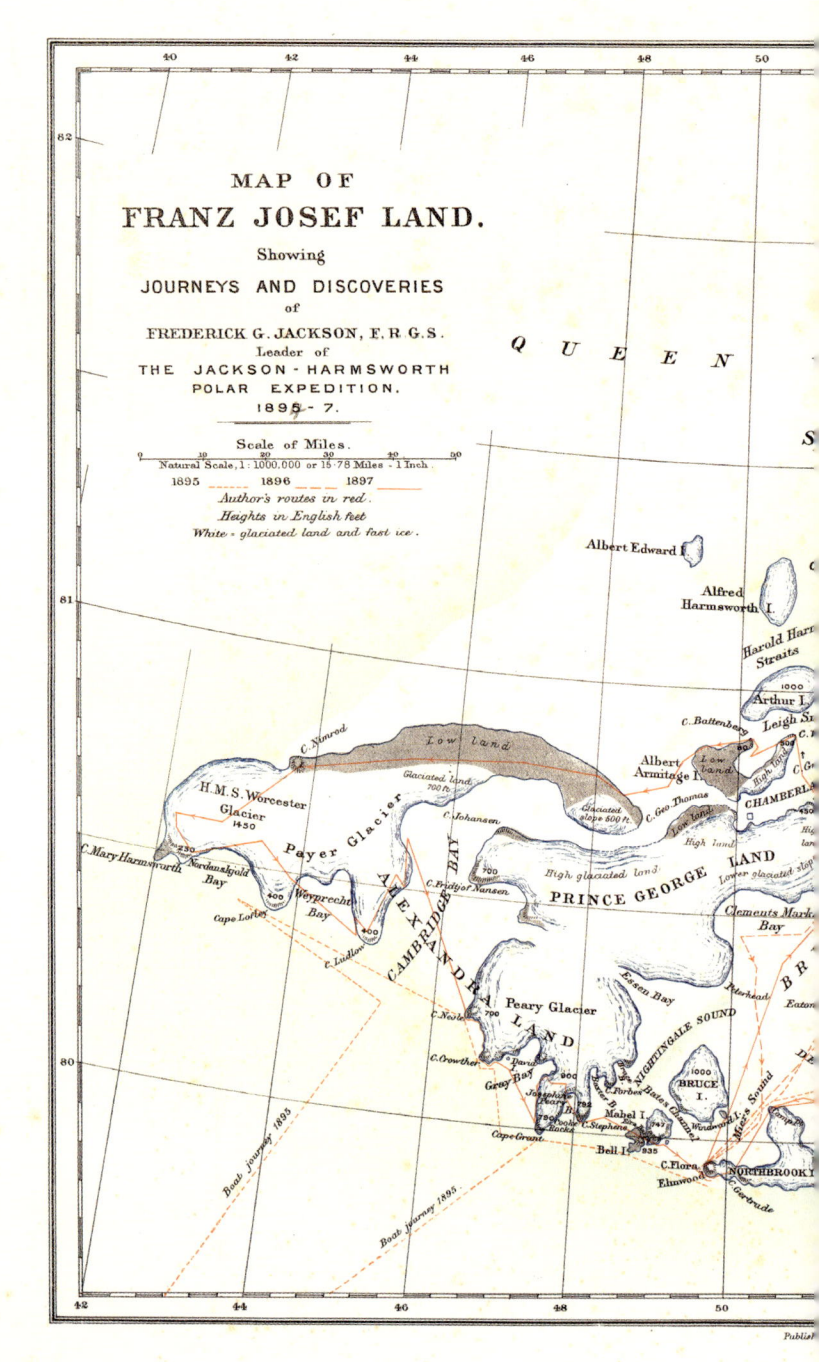

MAP OF
FRANZ JOSEF LAND.

Showing
JOURNEYS AND DISCOVERIES
of
FREDERICK G. JACKSON, F.R.G.S.
Leader of
THE JACKSON - HARMSWORTH
POLAR EXPEDITION.
1895 - 7.

Scale of Miles.
Natural Scale, 1 : 1,000,000 or 15·78 Miles = 1 Inch.
1895 ——— 1896 ——— 1897
Author's routes in red.
Heights in English feet.
White = glaciated land and fast ice.

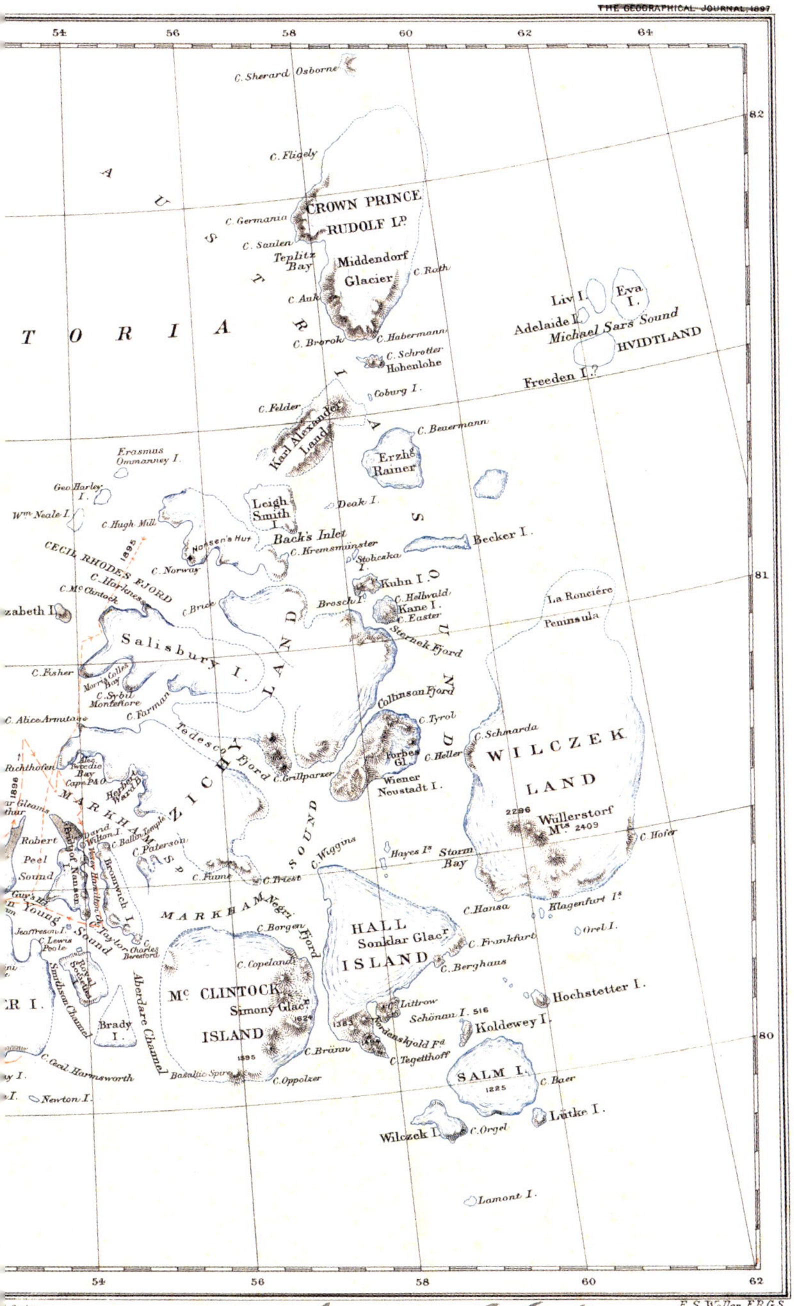

Frederick G. Jackson
Nov. 19. 1897

The last map in this chapter (figure 67) is not an elaborate map, but it captures the changing relationship between humans and polar bears. The map is also an example of how the polar bear came to be considered a natural part of scientific inquiry; it is a map that pulls at the heart strings of the beholder, despite its simplicity. The map, or rather the sketch, shows the outline of a polar bear lair, found at Cape Gertrude on the southernmost island of Franz Josef Land by the Jackson–Harmsworth expedition, on 3 February 1885.[42] More specifically, the lair was discovered by the dog Räwing, which also tried to attack the bear in it. Luckily for the dog, but less so for the bear, Räwing was saved by Jackson, who killed the bear with a single shot.

As a general rule, polar bears do not hibernate. However, female bears will make a lair when pregnant to give birth out of the wind and out of sight of predators. After giving birth, they spend the first few months of cubs' life in the lair. The polar bear that Räwing found was no exception. After having shot the mother, the crew extracted a cub from the lair: 'a white, soft, fluffy thing, hardly larger than a big cat'.[43] They returned to camp with the cub, intending to take her back to be exhibited at London Zoo.

The bear's lair was examined and mapped by the botanist and geologist Harry Fisher, and the result was the sketch map. The text next to the map details the physical particulars of the mother bear and her cub, including a detached note that the mother's womb 'had evidently quite recently been occupied'. To Jackson and his crew, polar bears could be dangerous, but they were also a curiosity and an object of study. The cub they had taken was named Gertie after Cape Gertrude, where she had been found. The crew fed the cub condensed milk and, after some time, meat. Later in the spring, they caught several more cubs, among them two siblings named Mabel and Benji. In naming the cubs and through actions such as photographing them, the expedition members anthropomorphized the bears. The bears were even kept in the same hut as the expedition's men. In Jackson's words, the cubs

A

Snow bank — Raising slated to known — 3½ feet snow above this

Breathing hole — Base

Horizontal section of hole

B

Transverse section at apex showing breathing hole 12 inches diam.

Transverse section at base. (the part occupied by mother & cub) on the 3rd Febr 1895.

♀ Bear (mother)
Length 6 ft 6 in from end of nose to tip of tail
Girth 4 ft 9 in round chest.
       4 ft 9 in   "   belly.
From nose end to a line between ears 15 in
Between base of the ears   11 in
Length of fore leg from top of shoulder to
    to end of claw   2 ft 9 in.
From elbow to claw end 10½ in.
"   heel to   "   "   15 in.
Thumb & little finger claws exserted 2⅜ in
Middle claw   3⅜ in (exserted)
Circumference of head just in front of
Ears 2 ft 3 in.
Weight 393 lbs
Teeth.
All these particulars taken 24 hours after death.

67 The botanist and geologist Harry Fisher, who was part of the Jackson-Harmsworth expedition to Franz Josef Land made this sketch map of a polar bear lair with a female bear and its newborn cub discovered by the dog Räwing.

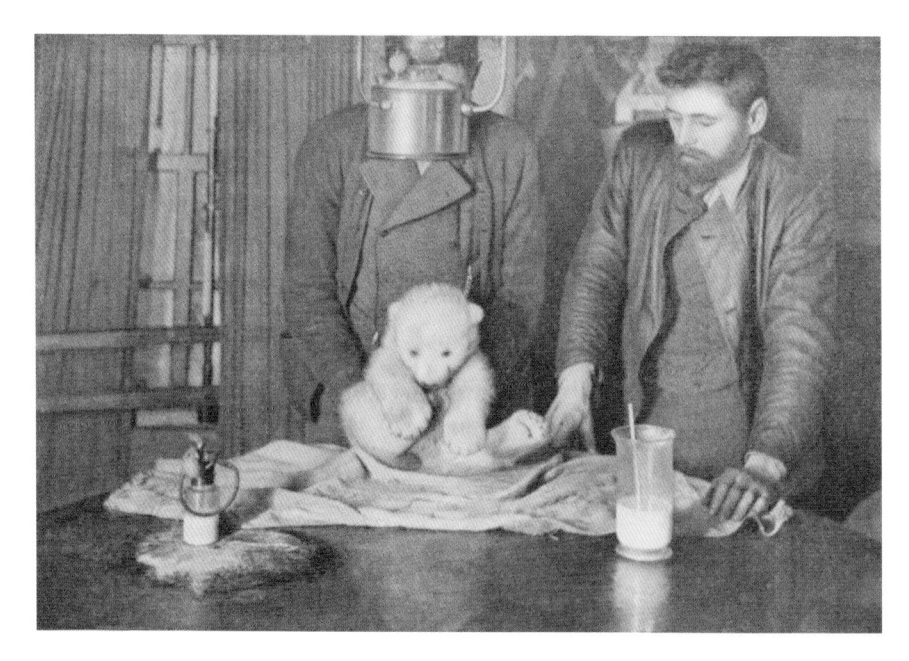

68 This photograph, entitled 'infant contentment after dinner', shows members of the Jackson–Harmsworth expedition feeding condensed milk to a polar bear cub. While their ambition was to bring the bears to London Zoo, the cubs all died. From *A Thousand Days in the Arctic*, 1899.

turned the hut 'into a perfect pandemonium of sound', and that seems hardly surprising.[44] Equally unsurprising is that all captured polar bears, including Gertie, died before long.

Today, the polar bear has become a symbol of human pollution and anthropogenic climate change. The polar bear is a maritime animal, largely dependent on the sea ice to hunt. When the ice melts, some groups of polar bears have a much harder time finding prey. That polar bears no longer appear on maps is more a result of changing cartographic conventions than climate change. Still, let us hope that coming generations can add a note to their maps of the north with the text 'here be white bears'.

FASCINATION, COHABITATION AND EXPLOITATION

A history of the mapping of the north that focuses on animals appears both similar to and refreshingly different from the account we have met earlier in the book. Differences between insider and outsider perspectives, the importance of economic incentives and the recurring exoticization of the north are familiar tropes by now. Viewing these themes, if not exactly through the eyes of the codfish at least with an eye to the importance of codfish, visualizes them anew.

With a focus on animals, it becomes clear that economic incentives and exoticization are neither mutually exclusive nor separable. To what extent a certain image or legend on a map should be understood as exotic or as showcasing natural resources is partly something that must be understood in relation to the world order and aesthetic conventions of a particular moment in time, and partly something that is determined by each individual viewer. For many of the maps discussed in this chapter, the inclusion of references to animals was both an acknowledgement of the commercial gains to be had in the north and, at the same time, an element of the map that functioned as a way to emphasize that this was a faraway and strangely different land.

There are enough differences between the north shown on maps of the Newfoundland coast and of Franz Josef Land that there might seem little point in grouping them together. However, all maps discussed in this chapter made an implicit or explicit claim about portraying the north, and all of them used animals as a part of that claim. Ideas about what and where the north is have varied greatly through history, yet where map-makers have portrayed animals as part of the north they have inadvertently tied these animals to geographical locations and included them as part of the natural landscape.

A

Hucusque extenditur regnum sueciæ

Insula Magnetum

B
Rubru panum hasta leuatum adorant

C
Starcaterus pugel sueticus

D

Tengillus rex scricfinorum

G

H

E

I

Arginus rex Helsingorum

Pele

Comutatio rerum

SCRICFINNIA

Ecclesia S. Andree

Olsby

I

Skrichi

Helsingaual

Silua longissima land stygia dicta

LACUS ALE
in quo innumere diuerse spes. pisc et auium

Helsigaly

Pirki

Pena

Vasmo

Laxastrem

Chim

Marturorum æprolo maxima multitudo

Lotra

F

K

Tehast

Hsum

Chalis

Ichia ula

LAPPIA ORIENTALIS

Pesca

L

Sala

Lergas

Trofel

Vorta

Karlabi

P

MARE BOTNICV

Psore

Tryla

Virtis

Pottra

Prefectus regisueciæ

Iokis

Viita

BOTNIA Orientalis

Vfro

Sauolax

D

A

Iomola ueti

CAR

kyro

Palio

kyla

Pageika

B

Monteiara Verga korihoim

Perkala

Piclmado

Iacusniger

K

Biri

H

Maxima hicmultitudo gentium

I

Holela Lacus

Vesikila

Surpesi

EPILOGUE

Throughout this book we have seen examples of how the making of a map has also been the making of a place. At times, the characteristics attributed to the north in maps reinforced a broader discourse on what the north was, such as the medieval maps of the climate zones that visualized an already inherently spatial argument. At other times, the nature of a particular kind of map seems to have taken the argument about what the north was one step further, such as when the format of the map or globe, with its unknown northern regions, incentivized map-makers to create a narrative of how geographical knowledge had expanded through the ages. Finally, maps have also presented novel views of the north, and most notably a view that nuances the theme of a barren, uninhabitable space. In fact, while I came to this inquiry expecting to find that maps of the north would mirror and reinforce the idea of an empty region, that assumption was largely unsubstantiated by the maps I studied. The idea of an empty north is part of how this area has been conceived; yet a focus on maps and mapping over a long period of time reveals just how incomplete that theme is.

Instead, the conceptions of the north that the maps in this book articulate are many and varied. That is to be expected since, in some ways, there is more that differentiates than unites the maps. Some maps were drawn by hand; others were printed in large editions. Some maps were lavish status symbols for their owners; others were mere doodles. Some maps were made by people with an intimate knowledge of the region they depicted; others were made by people who had never travelled farther than their own library. Some maps were intended to be symbolic in nature; others claim to show 'accurate and new' information about the geography of the northernmost parts of the world.

Nevertheless, all of the maps in this book have one crucial common denominator. Despite their many differences, they all present a spatial representation of the north as an idea, either in parts or as a contiguous whole. As such, the maps have contributed to the ongoing production of knowledge concerning the northernmost areas of our world. Simultaneously, through their subject matter and framing, these maps have helped define what and where 'the north' has been understood to be, and they have invested the north as a cultural and political idea with meaning.

Maps have mattered not least because, for much of recorded history, they have played an important role in informing people further south about the geography of the north. In perpetuating and providing new spatial demarcations to information in travel accounts, their impact outside the context of their production and use has been even greater than might have been expected. For that reason, maps need to be studied together with their texts and images, and understood as artefacts created at specific moments in time and in societies with varying standards regarding what was accepted as credible knowledge.

That said, there are also benefits to taking a broader outlook and following map-making over time. The very long history of the mapping of the north highlights patterns and breaks in how the north has been conceptualized in a way that narrower analysis of a particular moment in time could not have offered. For example, the long-term evolution of the polar bear from representing a danger to being an exotic and scientific curiosity, and then in the modern era having the status of an endangered species and symbolizing the dire effects of climate change, becomes clear in a diachronic and layered study.

The story of the mapping of the north would have changed significantly had this book looked further into the twentieth or even the twenty-first century. Surveying techniques such as aerial photography and, in more recent times, satellite imagery have made the north considerably less inaccessible. For the mapping of the northern ocean floors, the development of sonar technology has been crucial, though the majority of the oceans remain as yet unmapped. These technical advances have to some extent been promoted by idealistic notions of completing the 'world's map'. However, there are also very real economic incentives to continue the mapping of the north, such as the potential for exploiting oil and fishing resources. Knowledge of such commercial potential also sharpens political rivalries: witness how in the twentieth and twenty-first centuries tensions have repeatedly arisen over the control of northern territories.

While the north has become more accessible over time, so have we begun to realize that the riches are not endless. Olaus Magnus depicted a large pile of fish on the southern coast of Iceland on his map of the north, claiming that there was a 'vast number of fish heaped up for sale in a pile as high as the houses under the open sky'.[1] To the sixteenth-century observer of the map, this was meant to indicate that the north was rich in natural resources, ready to be made use of. Stepping outside this original context, the modern viewer is reminded of the controversies surrounding overfishing and of the wider perils to the natural environment that human societies bring. In the early decades of the twenty-first century, ideas about the end of time abound, yet we no longer fear for the northern hordes of Gog and Magog. We have moved from a state of anxiety about the north to anxiety for the north.

Olaus Magnus's pile of fish points to an opportunity offered by the study of maps. The maps in this book highlight how the north as a region and as an idea has been understood in different places and time periods, but their history also provides a framework for thinking about how we relate to the north today. Some of the themes that appear in medieval and early modern maps, such as the mythical islands and wondrous peoples of the north, have become outdated. Other ideas, such as the conception of a region removed from the concerns of the rest of the world, might still influence our way of thinking. Facing the already grave consequences of anthropogenic climate change, people in the north and south alike need to think carefully and critically about how we describe geography, climate and peoples from different parts of the world through texts, images and maps, and how such ideas have the potential to influence policy.

NOTES

INTRODUCTION

1. Olaus Magnus, *Carta Marina et descriptio septemtrionalium terrarum: ac mirabilium rerum in eis contentarum*, Venice, 1539. On Olaus Magnus, see Erling Sandmo, 'Dwellers of the Waves: Sea Monsters, Classical History, and Religion in Olaus Magnus's Carta Marina', *Norsk Geografisk Tidskrift – Norwegian Journal of Geography*, vol. 74, no. 4, 2020, pp. 237–49; Elena Balzamo, 'The Geopolitical Laplander: From Olaus Magnus to Johannes Schefferus', *Journal of Northern Studies*, vol. 8, no. 2, 2014, pp. 29–43; Leena Miekkavaara, 'Unknown Europe: The Mapping of the Northern Countries by Olaus Magnus in 1539', *Belgeo: Revue belge de géographie* 3–4, 2008, pp. 307–24; Karl Jakob Mauritz Ahlenius, *Olaus Magnus och hans framställning av Nordens geografi*, Almqvist & Wiksells, Uppsala, 1895.
2. Translations by the author unless otherwise stated.
3. Peter Davidson suggests that the image of Scandinavia became influential for broader conceptions of the north as an idea. Peter Davidson, *The Idea of North*, Reaktion Books, London, 2017, p. 172; Benedicte Gamborg Briså, 'Mapping the Expansion of the Known World in the North', *Norsk Geografisk Tidsskrift – Norwegian Journal of Geography*, vol. 74, no. 4, 2020, pp. 250–61, p. 252.
4. On reading maps historically, see Catherine Delano-Smith, Roger Kain and Katherine Parker, 'Maps and Mapping, History of', in *International Encyclopaedia of Human Geography*, ed. A. Kobayashi, 2nd edn, Elsevier, London, 2020, vol. 8, pp. 353–65; Matthew Edney, *Cartography: The Ideal and Its History*, University of Chicago Press, Chicago IL, 2019; J.B. Harley, *The New Nature of Maps: Essays in the History of Cartography*, ed. Paul Laxton, Johns Hopkins University Press, Baltimore MD and London, 2001.
5. For a similar argument regarding Arctic expeditions more generally, see Michael F. Robinson, *The Coldest Crucible: Arctic Exploration and American Culture*, University of Chicago Press, Chicago IL, 2006; and regarding travel accounts more broadly, see Stephen Greenblatt, *Marvelous Possessions: The Wonder of the New World*, University of Chicago Press, Chicago IL, 1992.

6. See Further Reading for a general survey of the field, and the relevant notes for information on specific maps and map-makers.
7. Pierre Salvadori, *Le Nord del la Renaissance: La carte, l'humanisme suédois et la genèse de l'Arctique*, Classiques Garnier, Paris, 2021; John Woitkowitz, 'Measuring and Mapping the Arctic: Cartography and the Legacies of Nineteenth-century Arctic Science', *Arctic Yearbook 2021*, https://arcticyearbook.com/images/yearbook/2021/Scholarly-Papers/7_AY2021_Woitkowitz.pdf, pp. 1–14; Michael Bravo, *North Pole*, Reaktion Books, London, 2019, ch. 2; Chet Van Duzer, 'The Mythic Geography of the Northern Polar Regions: Inventio fortunata and Buddhist Cosmology', *Culturas Populares. Revista Electrónica* 2, 2006, pp. 1–16.
8. Robert E. Peary, *The North Pole: Its Discovery in 1909 under the Auspices of the Peary Arctic Club*, F.A. Stokes, New York, 1910, p. 154. Peary was speaking to his times, as seen in contemporaneous projects like the International Map of the World initiated by the German geographer Albrecht Penk, or the 'Magnetic Crusade' headed by the English polymath Edward Sabine. See Peter Collier, 'Edward Sabine and the "Magnetic Crusade"', in Elri Liebenberg, Peter Collier and Zsolt Győző Török (eds), *History of Cartography: International Symposium of the ICA, 2012*, Springer, Berlin, 2014, pp. 309–24. On the drive to map the world, see Staffan Bergwik, *Terranauterna: Om människans dröm att upptäcka och kartlägga världen*, Norstedts, Stockholm, 2024.
9. For example, Thomas Reinertsen Berg recounts the difficulties of Fridtjof Nansen's survey of Greenland in 1898. Thomas Reinertsen Berg, *Kartornas Historia*, Lind, Stockholm, 2021, pp. 209–11.
10. Alfred Hiatt, *Terra Incognita: Mapping the Antipodes before 1600*, University of Chicago Press, Chicago IL, 2008, p. vi; see also *Mapping Uncertain Knowledge*, ed. Djoeke van Netten, special issue of the *Journal for the History of Knowledge* 5, 2024; Charlotta Forss, *The Old, the New and the Unknown: The Continents and the Making of Geographical Knowledge in Seventeenth-Century Sweden*, Illoinen tiede, Turku, 2018, especially pp. 107–11; Carla Lois, *Terrae incognitae: modos de pensar y mapear geografías desconocidas*, Eudeba, Buenos Aires, 2018.
11. Chet Van Duzer, 'Hic sunt dracones: The Geography and Cartography of Monsters', in Asa Mittman and Peter Dendle (eds), *The Ashgate Research Companion to Monsters and the Monstrous*, Ashgate Variorum, Farnham, 2012.
12. This is a rapidly changing field. Informative overviews can be found in Klaus Dodds and Jamie Woodward, *The Arctic: A Very Short Introduction*, Oxford University Press, Oxford, 2021; Klaus Dodds and Mark Nuttall, *The Arctic: What Everyone Needs to Know*, Oxford University Press, Oxford, 2019; Birgitta Evengård, Joan Nymand Larsen and Øyvind Paasche (eds), *The New Arctic*, Springer, Berlin, 2015.
13. Jen Hill, *White Horizon: The Arctic in the Nineteenth-Century British Imagination*, State University of New York Press, Albany NY, 2008, p. 3.
14. Dodds and Nuttall, *The Arctic*; Bernd Brunner, *Extreme North: A Cultural History*, W.W. Norton, New York, 2022, p. 69; Bernd Roling, Bernhard Schirg and Stefan Heinrich Bauhaus (eds), *Apotheosis of the North: The Swedish*

Appropriation of Classical Antiquity around the Baltic Sea and Beyond (1650 to 1800), De Gruyter, Berlin, 2017; Katharina N. Piechocki, 'Erroneous Mappings: Ptolemy and the Visualization of Europe's East', in Karen Newman and Jane Tylus (eds), *Early Modern Cultures of Translation*, University of Pennsylvania Press, Philadelphia PA, 2015; Charlotta Forss, 'Den tillfälliga omvärlden: Sverige, Norden och Europa i 1600-talets svenska tillfälleslitteratur', in Henrik Ågren (ed.), *Goda exempel: Värderingar och världsbild i tidigmodern svensk sakprosa och tillfällesdikt*, Historiska institutionen, Uppsala University, Uppsala, 2010.

15. For recent discussions about how to define 'the north', see Peder Roberts and Adrian Howkins, 'Introduction: The Problems of Polar History', in Adrian Howkins and Peder Roberts (eds), *The Cambridge History of the Polar Regions*, Cambridge University Press, Cambridge, 2023; Dodds and Woodward, *The Arctic*; Dodds and Nuttall, *The Arctic*.
16. Discussed by a wide array of authors, including Brunner, *Extreme North*; Nancy Campbell, *The Library of Ice: Readings from a Cold Climate*, Scribner, London, 2018; Sverker Sörlin, 'Cryo-history: Narratives of Ice and the Emerging Arctic Humanities', in Birgitta Evengård, Joan Nymand Larsen and Øyvind Paasche (eds), *The New Arctic*, Springer, Berlin, 2015; Sara Wheeler, *The Magnetic North: Travels in the Arctic*, Vintage Books, London, 2010.

THE UNKNOWN NORTH

1. The idea that medieval Europeans believed the Earth was flat has been thoroughly revised and debunked. Jeffrey Burton Russell, *Inventing the Flat Earth: Columbus and Modern Historians*, Praeger, New York, 1991.
2. Duane W. Roller, *Eratosthenes' Geography*, Princeton University Press, Princeton NJ, 2010, pp. 5–7; J. Lennart Berggren and Alexander Jones, *Ptolemy's Geography: An Annotated Translation of the Theoretical Chapters*, Princeton University Press, Princeton NJ, 2000, p. 20.
3. On Ptolemy, see Berggren and Jones, *Ptolemy's Geography*.
4. Incidentally, Surtsey did not exist in antiquity, being the result of a volcanic eruption in 1963.
5. Ptolemy, *Geographia*, book 1.1; O.A.W. Dilke, 'The Culmination of Greek Cartography in Ptolemy', in J.B. Harley and David Woodward (eds), *History of Cartography*, Volume 1: *Cartography in Prehistoric, Ancient, and Medieval Europe and the Mediterranean*, University of Chicago Press, Chicago IL and London, 1987, pp. 177–200, p. 183.
6. David J. Breeze and Alan Wilkins, 'Pytheas, Tacitus and Thule', *Britannia* 49, 2018, pp. 303–8.
7. The astronomer Geminos of Rhodes quotes Pytheas of Massalia on the length of the day in the north. See Geminos, *Geminos's Introduction to the Phenomena: A Translation and Study of a Hellenistic Survey of Astronomy*, ed. and trans. James Evans and J. Lennart Berggren, Princeton University Press, Princeton NJ, 2007, p. 162.
8. Breeze and Wilkins, 'Pytheas, Tacitus and Thule', p. 306.

9. Oxford, Bodleian Library, MS. Arab. c. 90. Yossef Rapoport and Emilie Savage-Smith provide a rich commentary, translation and facsimile; see Yossef Rapoport and Emilie Savage-Smith (eds and trans.), *An Eleventh-Century Egyptian Guide to the Universe: The Book of Curiosities*, Brill, Leiden, 2014.
10. Oxford, Bodleian Library, MS. Arab. c. 90, fols 24a–23b.
11. Rapoport and Savage-Smith, *An Eleventh-Century Egyptian Guide*, p. 417.
12. E.S. Kennedy and M.H. Kennedy, *Geographical Coordinates of Localities from Islamic Sources*, IGAIW, Frankfurt am Main, 1987, p. 353.
13. Yossef Rapoport, *Islamic Maps*, Bodleian Library Publishing, Oxford, 2020, p. 36.
14. For a discussion of the difficulty of identifying which people Ibn Faḍlān describes, see James E. Montgomery, 'Ibn Faḍlān and the Rūsiyyah', *Journal of Arabic and Islamic Studies* 3, 2017, pp. 1–25. On knowledge about the north of Europe in the Middle East in the medieval and early modern periods, see for example O.J. Tallgren-Tullio and A.M. Tallgren, 'La Finlande et les autres pays baltiques orientaux', *Studia Orientalia* 3, 1930.
15. Rapoport and Savage-Smith, *An Eleventh-Century Egyptian Guide*, p. 422.
16. On the early history of Gog and Magog, see Emeri van Donzel and Andrea Schmidt, *Gog and Magog in Early Eastern Christian and Islamic Sources: Sallam's Quest for Alexander's Wall*, Brill, Leiden, 2010.
17. King James Bible (1611), book of Ezekiel 38:9.
18. Anonymous, *The Alexander Poem*, cited in Van Donzel and Schmidt, *Gog and Magog*, p. 24.
19. Van Donzel and Schmidt, *Gog and Magog*, pp. 13–27.
20. Including another Islamic world map in the Bodleian Library collections: *Kitab al-Bad' wa-al-ta'rikh*, 977/1570, Oxford, Bodleian Library MS. Laud. Or. 317, fols 10v–11r. Mentioned in Van Donzel and Schmidt, *Gog and Magog*, p. 180.
21. On reading *mappae mundi*, see Matthew Boyd Goldie, 'Maps and the Medieval World at Large', in Raluca Radulescu and Sif Rikhardsdottir (eds), *The Routledge Companion to Medieval English Literature*, Routledge, New York, 2022; Felicitas Schmeider, 'Geographies of Salvation: How to Read Medieval Mappae Mundi', *Peregrinations: Journal of Medieval Art and Architecture*, vol. 6, no. 3, 2018, pp. 21–42; David Woodward, 'Medieval *Mappaemundi*', in Harley and Woodward (eds), *History of Cartography*, vol. 1, pp. 286–370.
22. For an overview of the extant versions of Higden's map, see Woodward, 'Medieval *Mappaemundi*'; on Higden's depiction of the north, see Michael Livingston, 'More Vinland Maps and Texts: Discovering the New World in Higden's Polychronicon', *Journal of Medieval History* 20, 2004, pp. 25–44; on the location of Paradise on medieval maps, see Alessandro Scafi, *Maps of Paradise*, University of Chicago Press, Chicago IL, 2013.
23. See Katharina N. Piechocki, 'Erroneous Mappings: Ptolemy and the Visualization of Europe's East', in Karen Newman and Jane Tylus (eds), *Early Modern Cultures of Translation*, University of Pennsylvania Press, Philadelphia PA, 2015; Charles Burnett and Zur Shalev (eds), *Ptolemy's Geography in the*

Renaissance, Warburg Institute, London, 2011; Patrick Gautier Dalché, 'The Reception of Ptolemy's *Geography* (End of the Fourteenth to Beginning of the Sixteenth Century)', in Harley and Woodward (eds), *History of Cartography*, vol. 1, pp. 285–364.

24. On conventions in mapping, see Mick Ashworth, *Why North Is Up: Map Conventions and Where They Came From*, Bodleian Library Publishing, Oxford, 2019.
25. Claudius Clavus, *Map of the North*, 1424–27, Nancy, Bibliothèque municipale, BM 441; Kirsten Andresen Seaver, 'Saxo Meets Ptolemy: Claudius Clavus and the "Nancy map"', *Norsk Geografisk Tidsskrift – Norwegian Journal of Geography*, vol. 67, no. 2, 2013, pp. 72–86.
26. For the debate on the relationship between the maps, see Seaver, 'Saxo Meets Ptolemy'.
27. On the mapping of Greenland, see Henrik Dupont, *Kortlægningen af Grønland*, Lindhardt og Ringhof, Copenhagen, 2022.
28. Benjamin Weiss, 'The *Geography* in Print: 1475–1530', in Charles Burnett and Zur Shalev (eds), *Ptolemy's Geography in the Renaissance*, Warburg Institute, London, 2011, pp. 91–120, p. 91; Karl-Heinz Meine, *Die Ulmer Geographia des Ptolemäus von 1842*, Anton H. Konrad Verlag, Weissenhorn, 1982.
29. Ptolemy, *Geographia*, Ulm, 1486, Oxford, Bodleian Library, MS. Arch. B b. 19; Debbie Hall, 'Old World to New World', in Hall (ed.), *Treasures from the Map Room: A Journey through the Bodleian Collections*, Bodleian Library Publishing, Oxford, 2016, p. 24.
30. On Macrobius's zonal map, see Alfred Hiatt, 'The Map of Macrobius before 1100', *Imago Mundi*, vol. 59, no. 2, pp. 149–76. For two critical perspectives on how to interpret the tradition of climate theory, see J.T. Olsson, 'The World in Arab Eyes: A Reassessment of the Climes in Medieval Islamic Scholarship', *Bulletin of the School of Oriental and African Studies* 77, 2014, pp. 487–508; Sara Miglietti, 'Climate Theory: An "Invented Tradition"?', in Charles Burnett and Pedro Mantas-España (eds), *Spreading Knowledge in a Changing World*, Córdoba University Press, Córdoba, 2019, pp. 205–24.
31. Incidentally, this is most likely where ice floes would appear in the spring, contributing to foggy weather and making seafaring difficult. I am grateful to Fredrik Charpentier Ljungqvist for drawing my attention to this.
32. Abraham Ortelius, 'Island', in *The Theatre of the Whole World*, John Norton & John Bill, London, 1606, p. 103. Ortelius's map of Iceland first appeared in 1585. See Peter van der Krogt (ed.), *Koeman's Atlantes Neerlandici*, vol. 3, part A, Hes & De Graaf, 't Goy-Houten, 2003, p. 93; Marcel van den Broecke, *Ortelius Atlas Maps: An Illustrated Guide*, HES Publishers, Westrenen, 1996, p. 212.
33. Andreas Bureus, *Orbis Arctoi Nova et Accurata Delineatio*, ed. and trans. Herman Richter and Wilhelm Norlind, C.W.K. Gleerups förlag, Lund, 1936, p. 20; on Bureus, see also Ulla Ehrensvärd, *History of the Nordic Map: From Myth to Reality*, John Nurminen Foundation, Helsinki, 2006.
34. On Neptune as a symbol of the bounty of the oceans, see Chet Van Duzer, *Frames that Speak: Cartouches on Early Modern Maps*, Brill, Leiden 2023, p. 204.

35. The colouring was likely done in the seventeenth or eighteenth century. I am grateful to Mats Höglund for advice on this point.
36. See Anne Christine Lien, 'Colouring Sovereignty: How Colour Helped Depict Territorial Claims to the Arctic in Northern Europe on Sixteenth to Nineteenth Century Maps', in Diana Lange and Benjamin van der Linde (eds), *Maps and Colours: A Complex Relationship*, Brill, Leiden, 2024, pp. 82–101. Lien discusses the colouring of Bureus's *Orbis Arctoi*, though not the partial copy kept in Uppsala University Library.
37. On de Capell Brooke, see Lennart Pettersson, 'Some Aspects on the Pictures of the North', *Nordlit* 23, 2008, pp. 251–72; James Marshall-Cornwall, *Geographic Journal*, vol. 144, no. 2, 1978, pp. 250–53.
38. Arthur de Capell Brooke, 'Map of Sweden, Norway and Lapland', in de Capell Brooke, *A Winter in Lapland and Sweden, with Various Observations Relating to Finmark and its Inhabitants, made during a Residence at Hammerfest, near the North Cape*, John Murray, London, 1826, p. vi.
39. Otto Julius Hagelstam, *Geografisk militairisk och statistisk karta öfver hela Sverige och Norrige*, [Stockholm?], 1820.
40. For example, de Capell Brooke succinctly translated the terms 'Jaure', 'Järwi, Jervi', 'Javri', 'Träsk or Traesk', 'Sio, Sjo, Sion', 'Söe, Söen', 'Vand' and 'Vatn' with the English word 'Lake'.
41. Peder Roberts and Lize-Marié van der Watt, 'On Past, Present and Future Arctic Expeditions', in Birgitta Evengård, Joan Nymand Larsen and Øyvind Paasche (eds), *The New Arctic*, Springer, Berlin, 2015; Mark Davies, *A Perambulating Paradox: British Travel Literature and the Image of Sweden c. 1770–1865*, University of Lund, Lund, 2000.
42. Charles-Joseph Minard, *Carte Figurative des pertes successives en hommes de l'armée française dans la Campagne de Russie 1812–13*, Paris, 1869.
43. On war, nature and culture in Minard's map, see Charles Travis, *Environment as a Weapon: Geographies, Histories and Literature*, Springer, Berlin, 2024, ch. 6; Alan Forrest and Peter Hicks, *The Cambridge History of the Napoleonic Wars*, Volume III: *Experience, Culture and Memory*, Cambridge University Press, Cambridge, 2022; on the time element in the map, see Kären Wigen and Caroline Winterer (eds), *Time in Maps: From the Age of Discovery to Our Digital Era*, University of Chicago Press, Chicago IL, 2020, p. x.
44. Abraham Ortelius, *Typus Orbis Terrarum* in *Theatrum Orbis Terrarum*, Antwerp, 1570; Marcel van den Broecke, *Ortelius Atlas Maps: An Illustrated Guide*, HES Publishers, Westrenen, 1996, p. 40.
45. I explore this point further in Charlotta Forss, *The Old, the New and the Unknown: The Continents and the Making of Geographical Knowledge in Seventeenth-Century Sweden*, Illoinen tiede, Turku, 2018. See also Carla Lois, *Terrae incognitae: modos de pensar y mapear geografías desconocidas*, Eudeba, Buenos Aires, 2018; Lucile Haguet, 'Specifying Ignorance in Eighteenth-Century Cartography, a Powerful Way to Promote the Geographer's Work: The Example of Jean-Baptiste d'Anville', in Cornel Zwierlein (ed.), *The Dark Side of Knowledge: Histories of Ignorance, 1400 to 1800*, Brill, Leiden, 2016;

Alfred Hiatt, *Terra Incognita: Mapping the Antipodes before 1600*, University of Chicago Press, Chicago IL, 2008.

46. It is rather ironic that, most likely, Columbus never set foot on mainland North America. See Eviatar Zerubavel, *Terra Cognita: The Mental Discovery of America*, Routledge, New York, 2003.

47. Gerhard Friedrich Müller, *Nouvelle Carte des Decouvertes Faites par des Vaisseaux Russiens aux Côtes Inconnues de l'Amerique Septentrionale Avec les Pais Adiacents*, St Petersburg, 1758; Alexey V. Postnikov, 'Russia', in Matthew H. Edney and Mary Sponberg Pedley (eds), *History of Cartography*, Volume 4: *Cartography in the European Enlightenment*, part 1, University of Chicago Press, Chicago IL, 2020; Paula van Gestel-van Het Schip, Joop Kaashoek, Jaap Molenaatr, Rob Poelijoe, Henk Schipper and Hans van der Zwan, *Maps in Books of Russia and Poland: Published in the Netherlands to 1800*, Hes & De Graaf, 't Goy-Houten, 2011, p. 315; Leo Bagrow, *A History of the Cartography of Russia*, vol. 2, Walker, Wolf Island, 1975, pp. 179–80, 196–7.

48. See Lisa Hellman, 'Drawing the Lines: Translation and Diplomacy in the Central Asian Borderlands', *Journal of the History of Ideas*, vol. 82, no. 3, 2021, pp. 485–501.

49. Jürgen Espenhorst, *Petermann's Planet*, Volume 1: *Guide to the Great Handatlases*, ed. George R. Crossman, Pangaea Verlag, Schwete, 2003, pp. 178ff.

50. Ibid., p. 266.

51. Ibid., p. 274.

52. Wigen and Winterer (eds), *Time in Maps*; Walter Goffart, *Historical Atlases: The First Five Hundred Years*, University of Chicago Press, Chicago IL, 2003.

53. Edward Quin, *An Historical Atlas, Containing Maps of the World at Twenty-One Different Periods*, Seeley & Burnside, London, 1830, p. 1.

54. Ibid., p. 2.

55. Africa followed instead a different, less linear trajectory, becoming 'unmapped' by European cartographers in the nineteenth century as a part of exploitative politics. See Petter Hellström, 'A New New World: Unmapping Africa in the Age of Reason', *Journal for the History of Knowledge* 5, 2024.

MAPS & FICTIONAL TRAVEL

1. On truth and travel writing, see for example Felicitas Schmeider, 'Old and New Land in the North and West: The North Atlantic on the Medieval Globe around 1500', *The Medieval Globe*, vol. 7, no. 1, 2021, pp. 131–51; Peter Hulme and Tim Youngs (eds), *The Cambridge Companion to Travel Writing*, Cambridge University Press, Cambridge, 2002.

2. There are different manuscript versions of the account, varying in emphasis and episodes included. See Glyn S. Burgess and Clara Strijbosch, *The Legend of St Brendan: A Critical Bibliography*, Royal Irish Academy, Dublin, 2000.

3. The dating of the *Navigatio* has long been a topic of debate. See J.S. Mackley, *The Legend of St. Brendan: A Comparative Study of the Latin and Anglo-Norman Versions*, Brill, Leiden, 2008, pp. 15–16.

4. Saint Brendan, *Navigatio Sancti Brendani Abbatis*, in *Lives of Saints including St.*

Brendan, Oxford, Bodleian Library, MS. Laud Misc. 173. See Dorothy Ann Bray, 'Allegory in the *Navigatio Sancti Brendani*', *Viator* 26, 1995, pp. 1–10; William Babcock, 'St Brendan's Explorations and Islands', *Geographical Review* 8, 1919, pp. 37–46, p. 40. On the origins of the idea of a whale posing as an island, see Chet Van Duzer, 'Floating Islands Seen at Sea: Myth and Reality', *Anuário do Centro de Estudios de História do Atlântico* 1, 2009, pp. 110–20: pp. 114–15.

5. See Scott D. Westrem, *The Hereford Map: A Transcription and Translation of the Legends with Commentary*, Brepols, Turnhout, 2001, p. 389; Robert D. Benedict, 'The Hereford Map and the Legend of St. Brandan', *Journal of the American Geographical Society of New York* 24, 1892, pp. 321–65, p. 344.
6. Abraham Ortelius, *Septentrionalium Regionum Descrip* in *Theatrum Orbis Terrarum*, Antwerp, 1570. See Peter van der Krogt (ed.), *Koeman's Atlantes Neerlandici* 3, part A, Hes & De Graaf, 't Goy-Houten, 2003, p. 49; Marcel van den Broecke, *Ortelius Atlas Maps: An Illustrated Guide*, HES Publishers, Westrenen, 1996, p. 211.
7. Guillaume Delisle, *Carte de la Barbarie de la Nigritie et de La Guinee*, Paris, 1707. Using the name 'St. Borondon' to refer to Saint Brendan.
8. Mackley, *The Legend of St. Brendan*, p. 43.
9. Olaus Magnus mentions Saint Brendan's adventure in *A Description of the Nordic Peoples, 1555*, vol. 3, ed. Peter Foote, The Hakluyt Society, London, 1998, book 21, ch. 26, p. 1109.
10. J.R.R. Tolkien, 'Fastitocalon', in *The Adventures of Tom Bombadil and Other Verses from the Red Book*, George Allen & Unwin, London, 1962.
11. Including Alfred, Lord Tennyson's poem '*The Kraken*' (1830) and Jules Verne's *Twenty Thousand Leagues under the Sea* (1870), as well as recent films such as *Clash of the Titans* (1981 and 2010) and *Pirates of the Caribbean* (2006).
12. Erich Pontoppidan, *Det første Forsøg paa Norges naturlige Historie*, vol. 2, Berlingske Arvingers Bogtrykkerie, Copenhagen, 1753, p. 340.
13. Caspar Plautius, *Nova typis transacta navigatio novi orbis Indiae occidentalis*, [Linz], 1621; Plautius wrote under the pseudonym Honorius Philophonus. Joelle Weis, 'Historisch-kritische Analyse der *Nova Typis Transacta Navigatio novi Orbis Indiae occidentalis* (Linz 1621)', Master's thesis, University of Vienna, 2014.
14. On Nicolò Zen the Younger and his travel account, see Elizabeth Horodowich, 'Nicolò Zen and the Virtual Exploration of the New World', *Renaissance Quarterly*, vol. 63, no. 3, 2014, pp. 841–77. The account was published together with another travel account by a member of the Zen family: Nicolò Zen, *De i commentarii del viaggio in Persia ... et dello Scoprimento dell' Isole Frislanda, Eslanda, Engrouelanda, Estotilanda, & Icaria, fatto sotto il Polo Artico, da due Fratelli Zeni*, Francesco Marcolini, Venice, 1558; an English edition can be found in Fred W. Lucas, *The Annals of the Voyages of the Brothers Nicolò and Antonio Zeno*, H. Stevens Son and Stiles, London, 1898.
15. See, for example, Ayesha Ramachandran, *The Worldmakers: Global Imagining in Early Modern Europe*, Chicago University Press, Chicago IL, 2015; Christine Johnson, *The German Discovery of the World*, University of Virginia Press,

Charlottesville VA and London, 2008; Anthony Grafton, *New Worlds, Ancient Texts: The Power of Tradition and the Shock of Discovery*, Belknap Press, Cambridge MA and London, 1992; Wilcomb Washburn, 'The Meaning of "Discovery" in the Fifteenth and Sixteenth Centuries', *American Historical Review*, vol. 68, no. 1, 1962, pp. 1–21.

16. For instance, Nicolò Zen seems to have drafted a will in the year 1400. See Edward Brooke-Hitching, *The Phantom Atlas: The Greatest Myths, Lies and Blunders on Maps*, Simon & Schuster, London, 2016, p. 243.
17. Horodowich, 'Nicolò Zen and the Virtual Exploration'.
18. Such as the Italian Pietro Querini, who was shipwrecked together with his crew off the Norwegian coast in the fifteenth century. Randi Lise Davenport, 'Dando crédito al Septentrión: Ricla y el naufragio de Pietro Querini en la isla de Røst', *Hipogrifo: Revista de Literatura y Cultura del Siglo de Oro*, vol. 7, no. 1, 2019, pp. 59–71; Nils M. Knutsen, *Mørkrets og kuldens rike: Tekster i tusen år om Nord-Norge og nordlendingene*, Cassiopeia förlag, Tromsø, 1994.
19. Lucas, *The Annals*, p. vii.
20. For an introduction to the myth, see M.C. Howatson, *The Oxford Companion to Classical Literature*, 3rd edn, Oxford University Press, Oxford, 2011, 'Argonauts'. On Apollonius Rhodius and his version of the tale, see Theodore D. Papanghelis and Antonios Regnakos, *Brill's Companion to Apollonius Rhodius*, 2nd edn, Brill, Leiden, 2008.
21. Francesca Cannella, '"The Heroes of the Fabulous History and the Invention Ennobled by Them": The Myth of the Argonauts between Visual Sources and Literary Inventio', *Music in Art*, vol. 40, no. 1–2, 2015, pp. 191–202. For an example of a scholar who has interpreted the myth as based on actual events, see C. Doumas, 'What Did the Argonauts Seek in Colchis?', *Hermathena* 150, 1991, pp. 31–41.
22. Olof Rudbeck, *Olaus Rudbecks Atlantica*, vol. 1, ed. Axel Nelson, Almqvist & Wiksell, Uppsala, 1937, pp. 326, 418–27. On Rudbeck, see David King, *Finding Atlantis: A True Story of Genius, Madness, and an Extraordinary Quest for a Lost World*, Harmony Books, New York, 2005; Gunnar Eriksson, *The Atlantic Vision: Olaus Rudbeck and Baroque Science*, Science History Publications, Canton MA, 1994.
23. On Olof Rudbeck's use of maps for conducting and legitimating research, see Charlotta Forss, 'Mapping Atlantis: Olof Rudbeck and the Use of Maps in Early Modern Scholarship', *Journal of the History of Ideas*, vol. 84, no. 2, 2023, 207–31.
24. Olof Rudbeck's treatise on the lymphatic system was published in 1653, although it built on research conducted in previous years. To Rudbeck's great consternation, he had to share the laurels for the discovery with the Danish scholar Thomas Bartholin. See Olof Rudbeck, *Nova Exercitatio Anatomica*, Eucharius Lauringer, Västerås, 1653.
25. Olof Rudbeck, 'A. Rudbecks skrifvelse till Rådet 12 Nov. 1677', in Gustaf Edvard Klemming (ed.), *Anteckningar om Rudbecks Atland*, P.A. Norstedt & söner, Stockholm, 1863.
26. Rudbeck, *Olaus Rudbecks Atlantica*, vol. 1, p. 422.

27. Abraham Ortelius, *Argonautica*, 1598. See Peter van der Krogt (ed.), *Koeman's Atlantes Neerlandici*, vol. 3, part A, Hes & De Graaf, 't Goy-Houten, 2003, p. 247; Marcel van den Broecke, *Ortelius Atlas Maps: An Illustrated Guide*, HES Publishers, Westrenen, 1996, p. 226.
28. Rudbeck, *Olaus Rudbecks Atlantica*, vol. 1, p. 326. An introduction by Georgius Hornius was included in Johannes Janssonius's collection of historical maps from 1653. It seems plausible that this is what caused Rudbeck to associate the map with Hornius. See Peter van der Krogt (ed.), *Koeman's Atlantes Neerlandici*, vol. 1, HES & De Graaf, 't Goy- Houten, 1997, pp. 209, 496. Rudbeck reproduced the map in Olof Rudbeck, *Taflor*, Uppsala, 1679.
29. William Poole and Kelsey Jackson Williams, 'A Swede in Restoration Oxford: Gothic Patriots, Swedish Books, English Scholars', *Lias*, vol. 39, no. 1, 2012, pp. 1–67.
30. In the sixteenth century *Inventio Fortunata* was believed to have been written by the Oxford Carmelite friar Nicholas of Lynn, although this is now considered an unlikely attribution. Chet Van Duzer, 'The Mythic Geography of the Northern Polar Regions: *Inventio fortunata* and Buddhist Cosmology', *Culturas Populares. Revista Electrónica* 2006, pp. 1–16.
31. Gerhard Mercator and heirs, *Septentrionalium Terrarum Descriptio*, in *Atlantis pars Altera*, Duisburg, 1595. On the similarities and differences between the two maps, see Edwin Okhuizen, 'De kaart van de Noordpol in Mercator's Atlas of 1595', *Caert-Thresoor*, vol. 13, no. 1, 1994, pp. 5–10; see also Van der Krogt (ed.), *Koeman's Atlantes Neerlandici*, vol. 1, pp. 50–54, 566.
32. While the geographic North Pole is presented as a magnetic rock, Mercator did also, along with many of his contemporaries, distinguish between the location of the magnetic and geographic poles. See Agustín Udías, 'Athanasius Kircher and Terrestrial Magnetism: The Magnetic Map', *Journal of Jesuit Studies*, vol. 7, no. 2, 2020, pp. 166–84. On knowledge about the Earth's magnetism in the nineteenth century, see Tim Fulford, Debbie Lee and Peter J. Kitson, *Literature, Science and Exploration in the Romantic Era: Bodies of Knowledge*, University of Cambridge Press, Cambridge, 2004.
33. Letter from Gerhard Mercator to John Dee, 20 April 1577, quoted in E.G.R. Taylor, 'A letter dated 1577 from Mercator to John Dee', *Imago Mundi*, vol. 13, no. 1, 1956, pp. 56–68, p. 60. See also Gerhard Mercator, *Atlas sive Cosmographicæ meditationes de fabrica mundi et fabricati figura*, ed. E.M. Ginger, Philip Smith et al., *Gerardus Mercator Duisburg, 1595: Atlas sive cosmographiae meditationes de fabrica mundi et fabricate figura*, The Lessing J. Rosenwald Collection Library of Congress, CD-ROM, Oakland, 2000, p. 162.
34. Gerhard Mercator, *Nova et Aucta Orbis Terrae Descriptio ad Usum Navigantium Emendate Accommodata*, 1569; citation from the English translation in 'Text and Translations of the Legends of the Original Chart of the World by Gerhard Mercator Issued in 1569', *International Hydrographics Review*, vol. 9, no. 2, 1932, pp. 7–45, p. 29.
35. The projection itself has proved important. However, Gaspar raises questions about the usefulness of Mercator's own map for contemporary navigators: Joaquim Alves Gaspar, 'Revisiting the Mercator World Map of 1569: An

Assessment of Navigational Accuracy', *Journal of Navigation* 69, 2016, pp. 1183–96. See also Karrow, 'Commentary'.
36. Martin Behaim, *Erdapfel* [Globe], Nuremburg, 1492; Johannes Ruysch, *Universalior cogniti orbis tabula*, Rome, 1507/8; Beau Riffenburgh, *Mapping the World: The Story of Cartography*, Carlton, London, 2014, pp. 42, 47; Thomas Suárez, *Shedding the Veil: Mapping the European Discovery of America and the World*, World Scientific Publishing, Singapore, 1992, pp. 40–41.
37. Ian Dear and Peter Kemp (eds), *The Oxford Companion to Ships and the Sea*, 2nd edn, Oxford University Press, Oxford, 2006, 'maelstrom'.
38. Here, Olaus Magnus builds on Jacob Ziegler, *Quae intus continentur Syria, Palestina, Arabia, Aegyptus, Schondia, Holmiae*, Strasbourg, 1532, fol. 101v; Olaus Magnus, *A Description of the Northern Peoples, 1555*, vol. 1, ed. Peter Foote, The Hakluyt Society, London, 1996, book 2, ch. 7, p. 101.
39. Elizabeth Sutton, 'Economics, Ethnography, and Empire: The Illustrated Travel Series of Cornelis Claesz, 1598–1603', vol. 1, thesis submitted in partial fulfilment of the requirements for the Ph.D. in Art History, University of Iowa, 2009, p. 100. See also Okhuizen, 'De kaart van de Noordpol in Mercator's Atlas of 1595', pp. 5–10.
40. Humphrey Gilbert, *A Discourse of a Discouerie for a New Passage to Cataia*, Henry Middleton for Richarde Ihones, London, 1576; Charles Henry Coote, 'Gilbert, Humphrey', in Leslie Stephen (ed.), *Dictionary of National Biography*, vol. 21, London, 1890. See also Benedicte Gamborg Briså, 'Mapping the Expansion of the Known World in the North', *Norsk Geografisk Tidsskrift – Norwegian Journal of Geography*, vol. 74, no. 4, 2020, pp. 250–61; Günter Schilder and Hans Kok, *Sailing across the World's Oceans: History & Catalogue of Dutch Charts Printed on Vellum 1580–1725*, Brill, Hes & De Graaf, Leiden, 2019, pp. 222–34.
41. Olof Rudbeck, *Olaus Rudbecks Atlantica*, vol. 2, ed. Axel Nelson, Almqvist & Wiksell, Uppsala, 1939.
42. Martin Frobisher, *The Three Voyages of Martin Frobisher: In Search of a Passage to Cathaia and India by the North-West, A.D. 1576–8*, ed. Richard Collinson, Cambridge University Press, Cambridge, 2010, p. 240.
43. Athanasius Kircher, *Mundus subterraneus in xii libros digestus*, vol. 1, Amsterdam, 1665, pp. 158–9. Much has been written about Kircher and his manifold scholarly interests. See, for example, Iva Lelková, 'Kircher, Athanasius', in Marco Sgarbi (ed.), *Encyclopedia of Renaissance Philosophy*, Springer, Berlin, 2022, pp. 1791–8; Paula Findlen (ed.), *Athanasius Kircher: The Last Man Who Knew Everything*, Routledge, New York, 2003.
44. Kircher, *Mundus subterraneus*, vol. 1, p. 174.
45. Mark A. Waddell, 'The World, As It Might Be: Iconography and Probabilism in the *Mundus subterraneus* of Athanasius Kircher', *Centaurus* 48, 2006, pp. 3–22.
46. Mary Shelley, *Frankenstein; or, The Modern Prometheus*, vol. 1, London, 1818, pp. 2–3; Jessica Richard, '"A paradise of my own creation": Frankenstein and the Improbable Romance of Polar Exploration', *Nineteenth-Century Contexts*, vol. 25, no. 4, 2003, pp. 295–314.

47. For instance, Gavin Francis retraces the paths of ancient and medieval travellers in *True North: Travels in Arctic Europe*, Polygon, Edinburgh, 2014.

ENCOUNTERS & EXPLORATION

1. Peter Davidson, *The Idea of North*, Reaktion Books, London, 2017.
2. William Phillip, 'To the right worshipfull, Sir Thomas Smith Knight', in Gerrit de Veer, *The Three Voyages of William Barents to the Arctic Regions (1594, 1595, and 1596)*, ed. Koolemans Beyen and Charles T. Beke, Cambridge University Press, Cambridge, 1876.
3. Fridtjof Nansen, *Farthest North*, vol. 1, Harper & Brothers, New York and London, 1897, p. 11.
4. Peter R. Martin, 'Indigenous Tales of the Beaufort Sea: Arctic Exploration and the Circulation of Geographical Knowledge', *Journal of Historical Geography* 2020, pp. 24–35; Andrew Stuhl, *Unfreezing the Arctic: Science, Colonialism and the Transformation of Inuit Lands*, University of Chicago Press, Chicago IL, 2017; G. Malcolm Lewis (ed.), *Cartographic Encounters: Perspectives on Native American Mapmaking and Map Use*, University of Chicago Press, Chicago IL, 1998.
5. On map decoration as information, see Lucile Haguet, 'Specifying Ignorance in Eighteenth-Century Cartography, a Powerful Way to Promote the Geographer's Work: The Example of Jean-Baptiste d'Anville', in Cornel Zwierlein (ed.), *The Dark Side of Knowledge: Histories of Ignorance, 1400 to 1800*, Brill, Leiden, 2016, p. 361.
6. Scott D. Westrem, *The Hereford Map: A Transcription and Translation of the Legends with Commentary*, Brepols, Turnhout, 2001, p. 187.
7. See Asa Mittman and Peter Dendle (eds), *The Ashgate Research Companion to Monsters and the Monstrous*, Ashgate Variorum, Farnham, 2012.
8. Westrem, *The Hereford Map*, p. 95.
9. Davidson, *The Idea of North*, p. 27.
10. Chet Van Duzer, '*Hic sunt dracones*: The Geography and Cartography of Monsters', in Mittman and Dendle (eds), *The Ashgate Research Companion to Monsters and the Monstrous*.
11. *The Vinland Sagas: The Icelandic Sagas about the First Documented Voyages across the North Atlantic*, ed. and trans. Keneva Kunz, Penguin Classics, London, 2008. On monstrous creatures and the Vinland Sagas, see Kirsten A. Seaver, '"Pygmies" of the Far North', *Journal of World History*, vol. 19, no. 1, 2008, pp. 63–87.
12. *Eirik the Red's Saga*, in *The Vinland Sagas*, p. 48. Other Icelandic sagas located mythical people in other parts of the world.
13. Henrik Dupont, *Kortlægningen af Grønland*, Lindhardt og Ringhof, Copenhagen, 2022; Andrew Wawn (ed.), *Northern Antiquity: The Post-medieval Reception of Edda and Saga*, Hisarlik Press, Enfield Lock, 1994. On the political uses made of the Old Norse sagas over time, see Bernd Brunner, *Extreme North: A Cultural History*, W.W. Norton, New York, 2022. The question of whether the Norse themselves made maps is discussed by Kirsten A.

Seaver, *Maps, Myths, and Men: The Story of the Vinland Map*, Stanford University Press, Redwood City CA, 2004.
14. Seaver, '"Pygmies" of the Far North', pp. 63–87; Karl Jakob Mauritz Ahlenius, *Olaus Magnus och hans framställning av Nordens geografi*, Almqvist & Wiksells, Uppsala, 1895, pp. 88–102.
15. Olaus Magnus, *Historia de gentibus septentrionalibus*, Rome, 1555. In English translation: Olaus Magnus, *A Description of the Northern Peoples, 1555*, vol. 1, ed. Peter Foote, The Hakluyt Society, London, 1996, book 2, ch. 11, p. 106.
16. Willem Blaeu, *Regiones Sub Polo Arctico*, Amsterdam, 1640; Peter van der Krogt (ed.), *Koeman's Atlantes Neerlandici*, vol. 2, Hes & De Graaf, 't Goy-Houten, 2000, p. 485. The cartouche and much of the geographical presentation also appear in Johannes Janssonius's *Nova et Accurata Poli Arctici*, Amsterdam, 1638. On the significance of cartouches in early modern mapping, see Chet Van Duzer, *Frames That Speak: Cartouches on Early Modern Maps*, Brill, Leiden, 2023.
17. On *cartes-à-figures* as a genre, see Günter Schilder, *Monumenta Cartographica Neerlandica*, vol. VI, Canaletto, Alphen aan den Rijn, 2000. On costume albums, see Ulinka Rublack, *Dressing Up: Cultural Identity in Renaissance Europe*, Oxford University Press, Oxford, 2011. On curiosity cabinets and collecting, see Paula Findlen, *Possessing Nature: Museums, Collecting, and Scientific Culture in Early Modern Italy*, University of California Press, Berkeley CA, 1996.
18. Sigtuna, Skokloster Slott, 10253_SKO, 10254_SKO, 10255_SKO, 10711_SKO; Charlotta Forss, 'En värld av kartor: Atlasverk och kartor i stormaktstidens Sverige', *Biblis* 63, 2013, pp. 18–27.
19. The map of America, Africa and Europe by De Wit, that of Asia by Valk. The ethnographic engravings first appeared in [Gerard Valk and Jan Luyten], *Costumes des quatre parties du monde*, [Amsterdam], 1670. See further Günter Schilder and Paula van Gestel-Van het Schip, *Wall Maps Printed in the Netherlands until 1800*, Brill, Hes & De Graaf, Leiden, forthcoming.
20. Johannes Schefferus, *Joannis Schefferi Argentoratensis. Lapponia Id est, Regionis Lapponum et gentis nova et verissima descriptio*, Frankfurt, 1673. See Linda Andersson Burnett, 'Translating Swedish Colonialism: Johannes Schefferus's *Lapponia* in Britain *c.*1674–1800', *Scandinavian Studies*, vol. 91, no. 1/2, 2019, pp. 134–62. On Sami and mapping, see Carl-Gösta Ojala and Jonas Moiné Nordin, 'Mapping Land and People in the North: Early Modern Colonial Expansion, Exploitation, and Knowledge', *Scandinavian Studies*, vol. 91, no. 1/2, 2019, pp. 98–133. On the exoticizing of the other, see for example Benjamin Schmidt, *Inventing Exoticism: Geography, Globalism, and Europe's Early Modern World*, University of Pennsylvania Press, Philadelphia PA, 2015.
21. Peter R. Martin, 'Indigenous Tales of the Beaufort Sea: Arctic Exploration and the Circulation of Geographical Knowledge', *Journal of Historical Geography* 67, 2020, pp. 24–35, p. 26.
22. Aaron Arrowsmith, *Map of the Countries Round the North Pole*, London, 1818. On Arrowsmith and the mapping of northern latitudes, see James Walker, 'Compiling "all the recent discoveries": Aaron Arrowsmith and

Mapping Western North America, 1790–1823', *IMCOS Journal* 147, 2016, pp. 25–40; Francis Herbert, 'A Cartobibliography (with locations of copies) of the [Aaron] Arrowsmith [sr] / [Edward] Stanford [sr] North Pole map, 1818–1937', *ACML Bulletin* 62, 1987, pp. 1–16.

23. On Arrowsmith and indigenous knowledge, see Barbara Belyea, 'Mapping the Marias: The Interface of Native and Scientific Cartographies', *Great Plains Quarterly*, vol. 17, no. 3/4, 1997, pp. 165–84.

24. Edward Stanford, *Map of the Countries Round the North Pole*, London, 1875.

25. Michael Bravo, *North Pole*, Reaktion Books, London, 2019; Adrian Howkins and Peder Roberts (eds), *The Cambridge History of the Polar Regions*, Cambridge University Press, Cambridge, 2023.

26. Peder Roberts and Eric Paglia, 'Science as National Belonging: The Construction of Svalbard as a Norwegian Space', *Social Studies of Science*, vol. 46, no. 6, 2016, pp. 894–911; Mark Monmonier, *From Squaw Tit to Whorehouse Meadow: How Maps Name, Claim, and Inflame*, University of Chicago Press, Chicago IL, 2006; Christian Jacob, *The Sovereign Map: Theoretical Approaches in Cartography throughout History*, University of Chicago Press, Chicago IL, 2006, esp. ch. 3.

27. William Scoresby, *An Account of the Arctic Regions*, vol. 1, Archibald Constable, Edinburgh, 1820, p. 94. The map appears in volume 2.

28. Scoresby, *An Account of the Arctic Regions*, vol. 1, p. 118. On the naming practices relating to Svalbard, see Norges Svalbard-og Ishavs-undersøkelser, *The Place-Names of Svalbard*, Kommisjon hos Jacob Dybwad, Oslo, 1942.

29. Ejnar Mikkelsen, *Two Against the Ice*, Travel Book Club, London, 1912, p. 95; Robert Peary, 'Report of R.E. Peary, C.E., U.S.N., on Work Done in the Arctic in 1898–1902', *Bulletin of the American Geographical Society*, vol. 35, no. 5, 1903, pp. 496–534; Henrik Dupont, *Kortlægningen af Grønland*, Lindhardt og Ringhof, Copenhagen, 2022, pp. 74–80.

30. On Arctic indigenous cultures, see for example Amber Lincoln, Jago Cooper and Jan Peter Laurens Loovers (eds), *Arctic: Culture and Climate*, British Museum and Thames & Hudson, London, 2020; Klaus Dodds and Mark Nuttall, *The Arctic: What Everyone Needs to Know*, Oxford University Press, Oxford, 2019. See also the twentieth-century artist Hans Ragnar Mathisen, who has made maps of Sápmi, the Sámi homeland, that make for an interesting contrast to Arrowsmith's view of the north: Hans Ragnar Mathisen, *Sábmi*, 1975.

31. On 'nature', 'culture' and 'environment' as historical concepts, see Alexandra Walsham, *The Reformation of the Landscape: Religion, Identity, and Memory in Early Modern Britain and Ireland*, Oxford University Press, Oxford, 2011; Sverker Sörlin and Paul Warde (eds), *Nature's End: History and the Environment*, Palgrave Macmillan, Basingstoke, 2009; Denis Cosgrove, *Social Formation and Symbolic Landscape*, University of Wisconsin Press, Madison WI, 1984.

32. The origin of the map has been the subject of debate. The map is now housed in the Pitt Rivers Museum in Oxford. See John R. Bockstoce, *Furs and Frontiers in the Far North: The Contest among Native and Foreign Nations*

for Control of the Intercontinental Bering Strait Fur Trade, Yale University Press, New Haven CT, 2009, pp. 268–73; Elena Okladnikova, 'Cartography in the Traditional African, American, Arctic, Australian, and Pacific Societies', in David Woodward and G. Malcolm Lewis (eds), *History of Cartography*, Volume 2, book 3: *Cartography in Prehistoric, Ancient, and Medieval Europe and the Mediterranean*, University of Chicago Press, Chicago IL, 1998, pp. 329–49.

33. Okladnikova, 'Cartography in the Traditional African, American, Arctic, Australian, and Pacific Societies'.
34. Gustav Holm, 'Etnologisk Skizze af Angmagsalikerne', in Gustav Holm (ed.), 'Den østgrønlandske expedition, udført i aarene 1883–85 under ledelse af G. Holm, anden del, text', *Meddelser om Grønland* 10, 1888, p. 143. See also Henrik Dupont, *Kortlægningen af Grønland*, Lindhardt og Ringhof, Copenhagen, 2022, pp. 67–8; Malcolm Lewis, 'Maps, Mapmaking, and Map Use by Native North Americans', in Woodward and Lewis (eds), *History of Cartography*, vol. 2, book 3, pp. 51–182, 168–9.
35. Holm, 'Etnologisk Skizze', p. 144; Gustav Holm, 'Tavle XXXXI. Traekaart, udskaarne af *Kuunit* fra *Umivik*', in Holm (ed.), 'Den østgrønlandske expedition, udført i aarene 1883–85 under ledelse af G. Holm, anden del, tavler', *Meddelser om Grønland* 10, 1888.
36. On seasonality in the Arctic, see Lincoln, Cooper and Loovers (eds), *Arctic Culture and Climate*. See also John Spink and D.W. Moodie, *Eskimo Maps from the Canadian Eastern Arctic*, *Cartographica* Monograph No. 5, University of Toronto Press, Toronto ON, 1972. On the long history of indigenous Arctic travel, see 'Pan Inuit Trails: Northwest Passage and the Construction of Inuit pan-Arctic Identities', www.paninuittrails.org/index.html (accessed 27 June 2024).
37. See Spink and Moodie, *Eskimo Maps from the Canadian Eastern Arctic*, pp. 4–7.
38. Malcolm Lewis ed., *Cartographic Encounters: Perspectives on Native American Mapmaking and Map Use*, University of Chicago Press, Chicago IL, p. 98; Spink and Moodie, *Eskimo Maps from the Canadian Eastern Arctic*, p. 7.
39. William Parry, *Journal of a Second Voyage for the Discovery of a North-west Passage from the Atlantic to the Pacific*, John Murray, London, 1824, p. 197.
40. See, for example, Russell A. Potter, 'Sir John Franklin and the Northwest Passage in Myth and Memory', in Adrian Howkins and Peder Roberts (eds), *The Cambridge History of the Polar Regions*, Cambridge University Press, Cambridge, 2023; Jen Hill, *White Horizon: The Arctic in the Nineteenth-century British Imagination*, State University of New York Press, Albany NY, 2008.
41. Letter from Francis Leopold McClintock to Lady Franklin, 'Fox. Ponds Bay, 3rd August/58', fol. 3v, Cambridge, Scott Polar Research Institute, MS 248/439/33. McClintock is using now outdated terminology to refer to the Inuit he met with.
42. R.T. Gould, *Chart showing the Vicinity of King William Island with the various positions in which relics of the Arctic Expedition under Sir John Franklin have been found*, London, 1927.
43. Potter, 'Sir John Franklin'.

ANIMALS ON NORTHERN MAPS

1. Scholars have paid increasing attention in recent years to the importance of animals in the history of the Arctic. See Amber Lincoln, Jago Cooper and Jan Peter Laurens Loovers (eds), *Arctic: Culture and Climate*, British Museum and Thames & Hudson, 2020; Sverker Sörlin, 'Comments by Sverker Sörlin, KTH Royal Institute of Technology, on: Andrew Stuhl, *Unfreezing the Arctic: Science, Colonialism, and the Transformation of Inuit Lands* (Chicago: University of Chicago Press, 2016)', *H-Environment Roundtable Reviews*, vol. 9, no. 1, 2019, pp. 11–18, p. 17. Fewer studies have centred on specific species of animals and asked where in the mapping processes they have had an impact; exceptions include Wilma George, *Animals and Maps*, University of California Press, Berkeley CA, 1969; and Djoeke van Netten's ongoing work on polar bears.
2. I am paraphrasing Jonathan Swift's famous passage here: 'So Geographers in *Afric*-Maps / With Savage-Pictures fill their Gaps; / And o'er unhabitable Downs / Place Elephants for want of Towns.' Jonathan Swift, *On Poetry: A Rapsody*, J. Huggonson, London, 1733. For a critique of the use of this quotation by map historians, see Matthew Edney, 'A Misunderstood Quatrain', in *Mapping as Process: A Blog on the Study of Mapping Processes: Production, Circulation, and Consumption*, 15 December 2018, www.mappingasprocess.net/blog/2018/12/15/a-misunderstood-quatrain (accessed 27 June 2024).
3. On human–animal relationships in Arctic cultures, see Lincoln, Cooper and Loovers (eds), *Arctic: Culture and Climate*; Paul Nadasdy, *Hunters and Bureaucrats: Power, Knowledge, and Aboriginal–State Relations in the Southwest Yukon*, UBC Press, Vancouver BC and Toronto ON, 2003.
4. Now at the Pitt Rivers Museum, Oxford.
5. George, *Animals and Maps*, p. 25.
6. Lincoln, Cooper and Loovers (eds), *Arctic: Culture and Climate*.
7. Robert E. Peary, *The North Pole: Its Discovery in 1909 under the Auspices of the Peary Arctic Club*, F.A. Stokes, New York, 1910, p. 5.
8. On Inuit knowledge and dress, see Beverly Lemire, 'Shirts and Snowshoes: Imperial Agendas and Indigenous Agency in Globalizing North America, *c*. 1660–1800', in Beverly Lemire and Giorgio Riello (eds), *Dressing Global Bodies: The Political Power of Dress in World History*, Routledge, Abingdon, 2020, pp. 65–84; Karl-Gunnar Norén, *Polarfararnas kläder: på liv och död*, Nielsen & Norén, Stockholm, 2018.
9. Diego Ribero, *Carta universal en que se contiene todo lo que del mundo se ha descubierto fasta agora*, Seville, 1529; original in the Vatican Library, Rome, reproduced here from facsimile by W. Griggs, London, [1887?]; quoted in K.J. Rankin and Poul Holm, 'Cartographical Perspectives on the Evolution of Fisheries in Newfoundland's Grand Banks Area and Adjacent North Atlantic Waters in the Sixteenth and Seventeenth Centuries', *Terrae Incognitae*, vol. 51, no. 3, 2019, pp. 190–218, p. 199. The name 'Cortereal' is a reference to a family of Portuguese explorers whose early journeys in the North Atlantic contributed significantly to what was known about this part of the world.

10. Rankin and Holm, 'Cartographical Perspectives', p. 194.
11. Kirsten A. Seaver, '"A very common and usuall trade": The Relationship between Cartographic Perceptions and "Fishing" in the Davis Strait *circa* 1500–1550', *British Library Journal*, vol. 22, no. 1, 1996, pp. 1–26, p. 5. See also Gregory McIntosh, 'The Piri Reis Map of 1528: A Comparative Study with Other Maps of the Time', *Mediterranea – ricerche storiche*, vol. 12, no. 34, 2015, pp. 303–18, p. 313.
12. Poul Holm et al., 'The North Atlantic Fish Revolution (ca. AD 1500)', *Quaternary Research* 2022, pp. 92–106.
13. Samuel de Champlain, *Carte geographiqve de la Novvelle Franse*, in *Les Voyages du Sieur de Champlain*, Paris, 1613.
14. Stephanie Pettigrew and Elizabeth Mancke, 'European Expansion and the Contested North Atlantic', *Terrae Incognitae*, vol. 50, no. 1, 2018, pp. 15–34; see also Seaver, '"A very common and usuall trade"'.
15. Rankin and Holm, 'Cartographical Perspectives', p. 214.
16. On Champlain as a map-maker, see David Buisseret, 'The Cartographic Technique of Samuel de Champlain', *Imago Mundi*, vol. 61, no. 2, 2009, pp. 256–9; Conrad E. Heidenreich and Edward H. Dahl, 'Samuel de Champlain's Cartography, 1602–32', in Raymonde Litalien and Denis Vagueois (eds), *Champlain: The Birth of French America*, trans. Käthe Roth, McGill–Queen's University Press and Les éditions de Septentrion, Montreal QC, 2004, pp. 312–32; Seymour I. Schwartz, '1500–1800', in Patricia Egan and Reginald Gay (eds), *The Mapping of America*, Harry N. Abrams, New York, 1980, pp. 85–93.
17. Thomas Jefferys, Emanuel Bowen and John Gibson, *An Accurate Map of North America*, in *The American Atlas*, Robert Sayer & John Bennett, London, 1776; J.B. Harley, 'The Bankruptcy of Thomas Jefferys: An Episode in the Economic History of Eighteenth century Map-making', *Imago Mundi*, vol. 20, no. 1, pp. 27–48, p. 38.
18. Fred Anderson, *Crucible of War: The Seven Years' War and the Fate of the Empire in British North America, 1754–1766*, Faber & Faber, London, 2000, pp. 483–4, 503–6.
19. Petter Dass, *Nordlands beskrivelse/Nordlands Trompet*, Copenhagen and Bergen, 1739. Thomas Reinertsen Berg, *Kartornas Historia*, Lind, Stockholm, 2021, pp. 270–74.
20. While the name *Pristis* refers to a genus of sawfish, which would not be found anywhere near the North Sea, *Physeter* is the genus of whales that include the sperm whale, a pelagic mammal that can be found off the coasts of Norway. Olaus Magnus, *Historia de gentibus septentrionalibus*, Rome, 1555; Olaus Magnus, *A Description of the Northern Peoples, 1555*, vol. 3, ed. Peter Foote, The Hakluyt Society, London, 1998, book 21, ch. 6, p. 1087.
21. See Olaus Magnus, *A Description of the Northern Peoples*, book 21, chs 13, 15, 20.
22. Elizabeth Ingalls, *Whaling Prints in the Francis B. Lothrop Collection*, Peabody Museum of Salem, Salem MA, 1987, p. 224. Similarly, a walrus on Jodocus Hondius's wall map of Europe (1595) is labelled 'monster'. See also

Erling Sandmo, 'Dwellers of the Waves: Sea Monsters, Classical History, and Religion in Olaus Magnus's Carta Marina', *Norsk Geografisk Tidskrift – Norwegian Journal of Geography*, vol. 74, no. 4, 2020, pp. 237–49; Chet Van Duzer, '*Hic sunt dracones*: The Geography and Cartography of Monsters', in Asa Mittman and Peter Dendle (eds), *The Ashgate Research Companion to Monsters and the Monstrous*, Ashgate Variorum, Farnham, 2012, p. 434.

23. Ingalls, *Whaling Prints*, p. xix. For an example, see William van der Gouwen, *Een walvisch, land 70 voeten, gestrandt op de Hollandtse zee-kust 1598*, Amsterdam, 1679.
24. Johan Baptist Homann, *Historia Animantium Marinorum Iconographica & Quidem Repraesentatio Ejus Specialis I, in qua Balaenarum Species … exhibentur*, [Nuremberg], [1752].
25. Ingalls, *Whaling Prints*, p. xviii.
26. Henricus Hondius, *Poli Arctici, et Circumiacentium Terrarum Descriptio Novissima* in *Atlas or a Geographicke description of the Regions, Countries and Kingdomes of the world*, Amsterdam, 1636. See Peter van der Krogt (ed.), *Koeman's Atlantes Neerlandici*, vol. 1, Hes Publishers, 't Goy-Houten, 1997, pp. 199–200, 566.
27. John R. Bockstoce, *Whales, Ice, and Men: The History of Whaling in the Western Arctic*, University of Washington Press, Seattle WA, 1986, pp. 13–14, p. 13.
28. Ingalls, *Whaling Prints*, p. xvi.
29. Heinrich Berghaus, *Physikalischer Atlas*, Gotha, 1838–48.
30. David Livingstone and Charles Withers, *Geographies of Nineteenth-century Science*, University of Chicago Press, Chicago IL, 2011, p. 442.
31. Andrew Stuhl, *Unfreezing the Arctic: Science, Colonialism and the Transformation of Inuit Lands*, University of Chicago Press, Chicago IL, 2017.
32. Paolo Chiesa, 'Marckalada: The First Mention of America in the Mediterranean Area (*c.* 1340)', *Terrae Incognitae*, vol. 53, no. 2, 2021, pp. 88–106.
33. Bartholomaeus Anglicus, cited in Robert Steele (ed. and trans.), *Mediaeval Lore from Bartholomew Anglicus*, Chatto & Windus, London, 1924, p. 101. Note that many versions of this phrase appear in medieval works on geography, not least that of Adam of Bremen. Michael Engelhard, 'Here Be White Bears', *Hakai Magazine*, 30 May 2017, https://hakaimagazine.com/features/here-be-white-bears/?xid=PS_smithsonian (accessed 27 June 2024). See also George, *Animals and Maps*, pp. 109–11.
34. Paolo Chiesa, 'Marckalada: The First Mention of America in the Mediterranean Area (*c.* 1340)', *Terrae Incognitae*, vol. 53, no. 2, 2021, pp. 88–106, p. 104; Theodore Ernest Hamy, *Études Historiques et Géographiques*, Ernest Leroux, Paris, 1896, p. 420.
35. Chet Van Duzer, *Martin Waldseemüller's 'Carta marina' of 1516: Study and Transcription of the Long Legends*, Springer Open, Cham, 2020, p. 63; Chet Van Duzer, 'Cartographic Invention: The Southern Continent on Vatican MS Urb. Lat. 274, Folios 73v–74r (*c.* 1530)', *Imago Mundi*, vol. 59, no. 2, 2007, pp. 193–222, p. 194.
36. See Monica Herrera-Casais, 'The 1413–14 Sea Chart of Aḥmad al-Ṭanjī', in E. Calvo, M. Comes, R. Puig and M. Rius (eds), *A Shared Legacy: Islamic Science*

East and West. Homage to Prof. J. M. Millàs Vallicrosa, Universitat de Barcelona, Barcelona, 2008, pp. 283–307, p. 293.
37. George, *Animals and Maps*, p. 91.
38. Gerrit de Veer, *The Three Voyages of William Barents to the Arctic Regions (1594, 1595, and 1596)*, ed. Koolemans Beyen and Charles T. Beke, Cambridge University Press, Cambridge, 1876. On Dutch early modern mapping of the Arctic, see Günter Schilder, *Monumenta Cartographica Neerlandica*, vol. IX, Hes & De Graaf, 't Goy-Houten, 2013, pp. 353–93.
39. De Veer, *The Three Voyages of William Barents*, p. 63.
40. Arthur Montefiore, 'The Jackson–Harmsworth North Polar Expedition: An Account of Its First Winter and of Some Discoveries in Franz Josef Land', *Geographical Journal*, vol. 6, no. 6, 1985, pp. 499–519, p. 511.
41. Ibid., p. 513. Cambridge, Scott Polar Research Institute, Smith, British Exploring Expedition, 1881–1882, MS 301/32, fol. 3r. See also Fredrick G. Jackson, 'Three Years' Exploration in Franz Josef Land', *Geographical Journal*, vol. 11, no. 2, 1989, pp. 112–38, p. 117.
42. Harry Fisher, 'Particulars of Polar Bear and its lair shot in Zemlya Frantsa-Iosifa on 3 February 1895', Cambridge, Scott Polar Research Institute, Harry Fisher collection, MS 287/28/2.
43. Jackson described the incidence in his journal: Frederick G. Jackson, *A Thousand Days in the Arctic*, Cambridge University Press, Cambridge, 1899, vol. 1, p. 163.
44. Ibid., p. 192.

EPILOGUE

1. Olaus Magnus, *Carta Marina et descriptio septemtrionalium terrarum: ac mirabilium rerum in eis contentarum*, Venice, 1539, quoted in *The Commentary by Olaus Magnus to the Map of the Scandinavian Countries*, Uppsala University Library, Uppsala, 1997.

FURTHER READING

This section provides guidance on research about the history of maps and mapping concerning the northernmost parts of the world, focusing on English-language books. This is a rich field, and there are many books and innumerable articles not mentioned here. However it is my hope that the essay will provide a useful starting point for further investigations. In addition, the endnotes contain references and commentary on research relating to individual maps, mapmakers and specific debates with which the book engages.

The first place to start for anyone interested in the history of maps and mapping is the six-volume series *The History of Cartography* published by University of Chicago Press. The first volume was published in 1987 under the editorship of J.B. Harley and David Woodward. At the time of writing, five volumes have been published, with one more on its way. The series is available online open-access: https://press.uchicago.edu/books/HOC/index.html. Numerous chapters are of relevance for the topic of this book.

In volume 2, book 3 (ed. David Woodward and Malcolm Lewis, 1998), Malcolm Lewis's chapter entitled 'Maps, Mapmaking, and Map Use by Native North Americans' deals with indigenous mapping in North America. In the chapter 'Cartography in the Traditional African, American, Arctic, Australian, and Pacific Societies', Elena Okladnikova writes about mapping in the north of Asia.

Volume 3 of *The History of Cartography* (ed. David Woodward, 2007) deals with maps and map-making in the European Renaissance. Of particular interest here are Felipe Fernández-Armesto's chapter in part 1, 'Maps and Exploration in the Sixteenth and Early Seventeenth Centuries', on Renaissance maps and exploration. In part 2 of the volume, take special note of William R. Mead's chapter 'Scandinavian Renaissance Cartography on the Nordic Region and L.A. Goldenberg, 'Russian Cartography to ca. 1700', on early modern map-making in Russia.

Volume 4 of *The History of Cartography* (ed. Matthew H. Edney and Mary Sponberg Pedley, 2019) focuses on the European Enlightenment. A large number of chapters in this volume are relevant for the history of the mapping of the north, but see in particular Olivier Chapus, 'North, Magnetic and True', Glyndwr Williams, 'Nothwest Passage', Henrik Dupont, 'Denmark and Norway', Jan Strang, 'Geographical Mapping in Sweden-Finland', in addition to Ulla Ehrensvärd's several chapters on Sweden and Alexey V. Postnikov and Nikolay N. Komedchikov's chapters on Russia. Furthermore, numerous other chapters in volume 3 and 4 touches on the cartography of the north made in different parts of Europe, for example several chapters on marine charting, on globes, atlases and world maps, and on the map trade.

Volume 6 of the *History of Cartography* (ed. Mark Monmonier, 2015) deals with mapping in the twentieth century. Numerous chapters on marine mapping and national developments are worth reading here, and of particular relevance is Derek Hayes chapter on the Arctic, 'Arctic, the'. Incidentally, Derek Hayes has also written an account of maps and the Arctic of a more popular, though informative, nature: *Historical Atlas of the Arctic* (University of Washington Press, 2003).

On indigenous maps and mapping, see Malcolm Lewis (ed.), *Cartographic Encounters: Perspectives on Native American Mapping and Map Use* (University of Chicago Press, 1998), and Daniel G. Cole and Imre Sutton (eds), *Mapping Native America: Cartographic Interactions between Indigenous Peoples, Government, and Academia* (CreateSpace, North Charleston SC, 2014). John Spink and D.W. Moodie, *Eskimo Maps from the Canadian Eastern Arctic*, Cartographica monograph no. 5 (University of Toronto Press, 1972) has an outdated nomenclature, yet is useful as a starting point for further inquiry. See also the interactive atlas project *Pan Inuit Trails: Northwest Passage and the Construction of Inuit pan-Arctic Identities*, www.paninuittrails.org/index.html.

For studies of early Dutch maritime cartography which also deal with the far north, see Günter Schilder, *Early Dutch Maritime Cartography: The North Holland School of Cartography (c. 1580–1620)* (Brill, Leiden, 2017), and Günter Schilder and Hans Kok, *Sailing Across the World's Oceans: History & Catalogue of Dutch Charts Printed on Vellum 1580–1725* (Brill, Hes & De Graaf, Leiden and Boston MA, 2019), especially chapter 5 on Dutch activities in the north-west Atlantic. For those interested in the details of the prolific early modern

Dutch map-making industry, Günter Schilder's series *Monumenta Cartographica Neerlandica* (Volumes I–VIII, Uitgeverij Canaletto/Repro Holland, Alphen aan den Rijn, 1986–2007; Volume IX: Hes & De Graaf, Houten, 2013) is of interest, as is the multivolume cartobibliography *Koeman's Atlantes Neerlandici*, edited by Peter van der Krogt (Volumes I–IV, Brill, Leiden, 1997–2012).

Several works on the mapping of North America provide useful context and cartobibilographic information. On the earlier history, see Lawrence B. Earl and Betty H. Kidd, *The Dawn of Arctic Cartography, Fourth Century to 1822* (Minister of Supply and Services, Ottawa, 1977), and Philip D. Burden's cartobibliographies of the mapping of North America in two volumes: *The Mapping of North America: A List of Printed Maps 1511–1670* (Raleigh Publications, Rickmansworth, 1996). On nineteenth-century developments, see Coolie Verner and Frances Woodward, *Explorers' Maps of the Canadian Arctic 1818–1860* (B.V. Gutsell, Department of Geography, York University, Toronto, 1972). See also the collected volume *Explorations in the History of Canadian Mapping*, edited by Barbara Farrell and Aileen Desbarats (Association of Canadian Map Libraries and Archives, Ottawa, 1988).

The history of maps and map-making in and of the Nordic region has been covered by several studies. Ulla Ehrensvärd, *History of the Nordic Map: From Myth to Reality* (John Nurminen Foundation, Helsinki, 2006), provides a contextualizing overview. William B. Ginsberg's cartobibliographies provide guidance to early printed maps of Norway, Scandinavia and the Arctic: see William B. Ginsberg, *Maps and Mapping of Norway, 1602–1855* (Septentrionalium Press, New York, 2009) and William B. Ginsberg, *Printed Maps of Scandinavia and the Arctic, 1482–1601* (Septentrionalium Press, New York, 2006). Recent updates to the field include the two special issues of *Norsk Geografisk Tidsskrift*, edited by Michael Jones: 'History of Cartography of the Nordic Countries', *Norsk Geografisk Tidsskrift* (*Norwegian Journal of Geography*), vol. 74, no. 4, 2020; vol. 75, no. 1, 2021.

For the history of Russian maps of the north – a topic which this book touches on briefly due to language and space constraints – Leo Bagrow's *A History of the Cartography of Russia*, in two volumes (Walker, Wolf Island, 1975), provides an important contribution. See also Fyodor A. Sibanov, *Studies in the History of Russian Cartography* (*Cartographica* monographa no. 14 and 15, University of Toronto Press, 1975), and Paula van Gestel-van Het Schip, Joop Kaashoek, Jaap Molenaatr, Rob Poelijoe, Henk Schipper and Hans van der Zwan, *Maps in Books of Russia and Poland: Published in the Netherlands to 1800* (Hes & De Graaf, 2011).

ACKNOWLEDGEMENTS

E VERY RESEARCH PROJECT is to some degree a collaborative effort, not least because every new work builds on the labour of so many previous investigations. Still, a book that from its conception takes a long-term perspective and a global outlook is, perhaps to a greater extent than many others, a collaborative product. I am grateful to all who have encouraged and assisted me. Any remaining errors are my own.

The map team at the Bodleian Library headed by Nick Millea has provided invaluable support, both through guidance to the map collections at the Bodleian Libraries and in sharing coffee breaks at the Headley Tea Room. At Uppsala University Library, Mats Höglund has helped me in identifying relevant maps and sharing his knowledge about the collections. At the Royal Library in Stockholm, Greger Bergvall and his colleagues have likewise been supportive, and the staff at the Scott Polar Research Institute Library in Cambridge have generously provided advice on the Institute's archival collections and library resources.

I am glad to have had the chance to present aspects of this work at seminars, conferences and workshops, among others at the Oxford Seminars in Cartography (TOSCA), the Maps and Society Lecture Series at the Warburg Institute, the Royal Institute of Technology in

Stockholm, Gothenburg University, the University of Hamburg and the International Conference on the History of Cartography (ICHC).

Several people have read parts of the manuscript at various stages. Thank you Fredrik Charpentier Ljungquist, Anna Derksen, Mari Eyice, Hanna Filipova, Kim Forss, Monica Gines, Mats Hallenberg, Lisa Hellman, Petter Hellström, Anna Knutsson, Otso Kortekangas, Carl Marklund, Ale Pålsson, Patrick van der Geest and Emily Winkler for your valuable comments. Thank you also to Colin Dupont, Gunlög Fur, Francis Herbert, Alfred Hiatt, Gwendolyne Knight, Martijn Storms, Yossef Rapoport, Paula van Gestel-van het Schip and Kären Wigen for discussing particular points with me. The late Erling Sandmo first prompted me to think about maps in relation to the idea of the north, and it is a great loss that he is no longer here to continue the discussion.

It has been a pleasure to work with the team at the Bodleian Library Publishing and illuminati books. It was Samuel Fanous who first asked if there might be a book waiting to be written on the topic I had originally thought of as a lecture. Later, Janet Phillips and Leanda Shrimpton encouraged and guided me through the work process. Thank you for your patience and help! Thank you also to the anonymous reviewer for your meticulous reading of my manuscript. Alan Shima has, once again, made an essential contribution through his excellent proofreading. A heartfelt thank you to the funding agencies behind this project, Axel and Margaret Ax:son Johnsons Stiftelse and Sven and Dagmar Saléns Stiftelse.

Finally, thank you to my family and friends in England and Sweden for sharing breaks with me. To Emil, thank you for your encouragements and support, there would not have been a book without you. To Axel and Ellen, I love you more than I love old maps. As you know, that says a lot.

I dedicate this book to my parents, Ulla Eriksson and Kim Forss, who kindled my love for travel, both in the real world and the armchair variety.

IMAGE SOURCES

1 Olaus Magnus, [*Carta Marina*], Antoine Lafréry, Rome, 1572. National Library of Sweden, Stockholm, KoB 1 ab. Detail from Olaus Magnus, *Carta Marina et descriptio septemtrionalium terrarum: ac mirabilium rerum in eis contentarum diligentissime elaborata*, Venice, 1539. Uppsala University Library, 88495 (alvin-record).
2 Andreas Bureus, *Orbis arctoi nova et accurata delineatio*, Stockholm, 1626. Uppsala University Library, 92658 (alvin-record)
3 François Étienne Musin, *HMS Erebus in the Ice*, oil on canvas, 1846. GL Archive/Alamy.
4 Claudius Ptolemy, *Totius Orbis Habitabilis Brevis Descriptio*, in *Geographia*, 1450–75, British Library/Bridgeman Images, MS. Harley 7182, fols 58v–59r.
5 [Rectangular world map], *Kitāb Gharā'ib al-funūn wa-mulaḥ al-'uyūn* ('The Book of Curiosities'), copied *c.* 1200. Oxford, Bodleian Library, MS. Arab. c. 90, fols 24a–23b.
6 Iskandar builds a wall against Gog and Magog, from the *Book of Divination* (*Fālnāma*) for Shah Ṭahmāsp, Qasvin, 1550–60. Chester Beatty, Dublin, CC BY 4.0., Per 395.1.
7 *Psalter mappa mundi*, England, *c.* 1265. British Library/Bridgeman Images, Add. MS. 28681, fol. 9r.
8 Ranulf Higden, *Mappa Mundi*, in *Polychronicon*, England, late 14th century. Oxford, Bodleian Library, MS. Tanner 170, fol. 15v.
9 [Ptolemaic world map], in *Ptolemaios Geographia*, Ulm, 1486. Oxford, Bodleian Library, Arch. B b.19.
10 Claudius Clavus, *Europe tabula XI*, in *Claudii Ptolemei cosmographiae libri VIII, latine versi a Jacobo Angelo Florentino*, 1424–27. Bibliothèque de Nancy/Limédia Galeries, Ms. 354, fols 184v–185r.
11 Nicolaus Germanus, [Map of northern Europe], in *Ptolemaios Geographia*, Ulm, 1482. Uppsala University Library, 8689 (alvin-record).
12 Macrobius, [Macrobian *mappa mundi*], in *De diuersis generibus musicorum*, Germany, late 10th century. Oxford, Bodleian Library, MS. D'Orville 77, fol. 100r.
13 Abraham Ortelius, *Iceland*, in *Theatrum orbis terrarum/The theatre of the whole world*, London, 1606 [1608?]. Oxford, Bodleian Library, Douce O subt. 15.
14 Hand-coloured detail of Andreas Bureus, *Orbis arctoi nova et accurata delineatio*, Stockholm, 1626. Uppsala University Library, 16861 (alvin-record).
15 Arthur de Capell Brooke, *Map of Sweden, Norway and Lapland*, in *A Winter in Lapland and Sweden*, J. Murray, London, 1827. Oxford, Bodleian Library, 310.S.1.

16 William Westall, *The author in his winter dress, as he travelled through Lapland*, in Arthur de Capell Brooke, *A Winter in Lapland and Sweden*, J. Murray, London, 1827. Oxford, Bodleian Library, 310.S.1.
17 Charles-Joseph Minard, *Carte Figurative des pertes successives en hommes de l'armée française dans la Campagne de Russie 1812–13*, Paris, 1869. Bibliothèque nationale de France, département Cartes et plans, GE DON-4182.
18 Abraham Ortelius, *Typvs Orbis Terrarvm*, in *Theatrum orbis terrarum/The theatre of the whole world*, London, 1606 [1608?]. Oxford, Bodleian Library, Douce O subt. 15.
19 Gerhard Friedrich Müller, *Nouvelle Carte des Decouvertes Faites par des Vaisseaux Russiens aux Côtes Inconnues de l'Amerique Septentrionale Avec les Pais Adiacents*, St Petersburg, 1758. Oxford, Bodleian Library, (E) F1:3 (1).
20 Hermann Berghaus and Adolf Stieler, *Nord-Polar-Karte*, in *Stielers Atlas of Modern Geography*, 9th ed., Justus Perthes, Gotha, 1911. Oxford, Bodleian Library, B1 b.67.
21 Hermann Haack and Adolf Stieler, *Nordpol*, in *Hand Atlas über alle Theile der Erde/ Stielers Handatlas*, 10th edn, Justus Perthes, Gotha, [1921–1925]. Oxford, Bodleian Library, previous shelfmark Y.2, Y.STI.
22 Edward Quin, *B.C. 2348. The Deluge*, in *A Historical Atlas*, London, 1830. Oxford, Bodleian Library, 2023 b.9.
23 Edward Quin, *A.D. 1828. End of the general peace*, in *A Historical Atlas*, London, 1830. Oxford, Bodleian Library, 2023 b.9.
24 *Nauigatio S. Brendani*, in *Saints' lives*, composite manuscript, 14th or early 15th century. Oxford, Bodleian Library, MS. Laud Misc. 173, fol. 129r.
25 Detail from *Nauigatio S. Brendani*, in *Saints' lives*, composite manuscript, 14th or early 15th century. Oxford, Bodleian Library, MS. Laud Misc. 173, fol. 115r.
26 Hereford *mappa mundi*, Hereford, *c*. 1290. Bridgeman Images/Hereford Cathedral.
27 Abraham Ortelius, *Septentrionalium Regionum Descrip.*, in *Theatrum oder Schawplatz des erdbodems*, 1572. University Library of Humboldt-Universität zu Berlin: 2018 C 31.
28 Detail from Olaus Magnus, *Carta Marina et descriptio septemtrionalium terrarum: ac mirabilium rerum in eis contentarum diligentissime elaborata anno dni 1539*, Venice, 1539. Uppsala University Library. 88495 (alvin-record).
29 [Caspar Plautius], *Nova typis transacta navigatio Novi orbis Indiae Occidentalis ... nunc primum e variis, scriptoribus in unum collecta ... Authore ... Honorio Philophono*, [Linz], 1621, plate 2. Oxford, Bodleian Library, fol. THETA 519.
30 Nicolo Zen the Younger, *Septentrionalium Partium Nova Tabula*, Girolamo Ruscelli, Venice, [1561–4]. Cornell University, PJ Mode Collection of Persuasive Cartography, 8548.
31 Olof Rudbeck, *Et Nos Homines*, in *Taflor*, [Uppsala], 1679. Oxford, Bodleian Library, Vet. C3 b.1, frontispiece.
32 Olof Rudbeck, *Tab IV. Fig 8. Tabula ex Geographia Antiqua Horny excerpta Argonautarum reditus ex Orpheo*, in *Taflor*, [Uppsala], 1679. Oxford, Bodleian Library, Vet. C3 b.1
33 Abraham Ortelius, *Argonavtica*, in *Parergon*, Antwerp, 1624. Wikimedia Commons.
34 Gerhard Mercator, *Septentrionalium Terrarum Descriptio*, in Rumold Mercator ed, *Atlas, siue Cosmographicæ meditationes de fabrica mundi et figura fabricati*, Duisburg, 1595. Oxford, Bodleian Library, Arch. B b.12.
35 Athanasius Kircher, *Poli Arctici Constitutio/ Poli Antarctici Constitutio*, in *Athanasii Kircheri e Soc. Jesus Mundus subterraneus in XII libros digestus* [*Mundus subterraneus*], Amsterdam, 1678. Oxford, Bodleian Library, Douce K 149, p. 170.
36 Athanasius Kircher, *Systema Ideale Qvo Exprimitur, Aquarum per Canales hydragogos subterraneos ex mari et in montium hydrophylacia protrusio, aquarumq. subterrestrium per pyragogos canales concoctus*, in *Athanasii Kircheri e Soc. Jesus Mundus subterraneus in XII libros digestus* [*Mundus subterraneus*], Amsterdam, 1678. Oxford, Bodleian Library, Douce K 149, opp. p. 186.

37 Frederic Edwin Church, *Aurora Borealis*, oil on canvas, 1865. Smithsonian American Art Museum, Gift of Eleanor Blodgett, 1911.4.1.
38 Guðbrandur Þorláksson, [Map of the North Atlantic Ocean], in Biørn Jonsen paa Skarsaa, *Grønlands Beskriffvelse*, 1669. Royal Library of Denmark, GKS 2881, fol. 9v.
39 Illustration from Olaus Magnus, *Historia de gentibus septentrionalibus*, Rome, 1555. Oxford, Bodleian Library, Douce M 723, p. 71.
40 Willem Blaeu, *Regiones Sub Polo Arctico*, in *Theatrum orbis terrarum, sive, Atlas novus*, Amsterdam, 1648. Oxford, Bodleian Library, Map Res. 23, following p. 4.
41 Frederick De Wit, Gerard Valk and J. De Ram, *Nova totius America tabula per J. de Ram*, Amsterdam, 1672. Skokloster Castle, Sigtuna, Sweden, 10255 SKO.
42 *Straet Davis en Hudson*, in Frederick De Wit, Gerard Valk and J. De Ram, *Nova totius America tabula per J. de Ram*, Amsterdam, 1672.
Skokloster Castle, Sigtuna, Sweden, 10255 SKO.
43 *Laplant*, in Frederick De Wit, Gerard Valk and J. De Ram, *Nova totius Europa tabula per J. de Ram*, Amsterdam, 1672. Skokloster Castle, Sigtuna, Sweden, 10254 SKO
44 Johannes Schefferus, *The History of Lapland: wherein are shewed the original, manners, habits, marriages, conjurations, &c. of that people*, Oxford, 1674. Oxford, Bodleian Library, Lawn d.55, title page.
45 Aaron Arrowsmith, *Map of the Countries Round the North Pole*, London, 1818. Oxford, Bodleian Library, (E) M1 (124).
46 Paul Goodhead, [Map of Arctic], in Amber Lincoln, Jago Cooper and Jan Peter Laurens Loovers eds., *Arctic Culture and Climate*, Thames and Hudson/British Museum Press, 2020, Fig 1. © The Trustees of the British Museum.
47 William Scoresby, *A chart of Spitzbergen or East Greenland comprising an original survey of the west coast*, in *An account of the Arctic regions with a history and description of the northern whale-fishery*, vol. 2, Edinburgh, 1820. Oxford, Bodleian Library, 8° O 100 BS.
48 Yupik map of Bering Strait on sealskin, second half of 19th century. Pitt Rivers Museum, University of Oxford, 1966.19.1.
49 Kuniit, [Map of eastern Greenland], *c.*1885. Facsimile from 2009, original in Greenland National Museum & Archives, Nuuk. Oxford, Bodleian Library, M4 d.1.
50 Iligliuk, *Eskimaux Chart No 1*, in William Parry, *Journal of a second voyage for the discovery of a north-west passage from the Atlantic to the Pacific*, London, 1824. Oxford, Bodleian Library, Dunston F 356, p. 196.
51 Sir John Ross, *Itmalick and Apelagliu, interviewed aboard Victory*, 1829–33. Scott Polar Research Institute, University of Cambridge, acc. Number 66/3/2.
52 R.T. Gould, *Chart showing the Vicinity of King William Island with the various positions in which relics of the Arctic Expedition under Sir John Franklin have been found*, London, 1927. Library & Archives Canada, e1076 1880.
53 *Nova Scotia and Newfoundland*, in [a volume of maps from R.M. Martin and J. & F. Tallis, *Illustrated atlas, and modern history of the World*, London & New York, 1851]. Oxford, Bodleian Library, B1 a.51.
54 Bow drill of walrus tusk with incised decoration, before 1826. Pitt Rivers Museum, University of Oxford, 1886.1.693.
55 Detail from Bartholomaeus Anglicus, *De proprietatibus rerum*, early 14th century, fol. 1r. Oxford, Bodleian Library, MS. Bodl. 965B.
56 *Peary searching the horizon for land*, in Robert E. Peary, *The North Pole: its discovery in 1909 under the auspices of the Peary Arctic Club*, Frederick A. Stokes, New York NY, 1910. Oxford, Bodleian Library, 2035 d.39.
57 Detail from Diego Ribero, *Carta universal en que se contiene todo lo que del mundo se ha descubierto fasta agora*, Seville, 1529. Facsimile by W. Griggs, London, [1887?],

IMAGE SOURCES 207

original in the Vatican Library, Rome. Library of Congress Geography and Map Division, Washington DC.
58 Samuel de Champlain, *Carte geographiqve de la Novvelle Franse*, in *Les Voyages du Sieur de Champlain*, Paris, 1613. Facsimile by A. Pilinski et Fils, 1880. Bibliothèque nationale de France, département Cartes et plans, GE C-9476.
59 Thomas Jeffreys, *An Accurate Map of North America. Describing and distinguishing the British and Spanish Dominions on the great Continent; According to the Definitive Treaty Concluded at Paris 10th Feby. 1763*, R. Sayer and J. Bennett, London, 1776. David Rumsey Map Collection.
60 Detail from Olaus Magnus, [*Carta Marina*], Antoine Lafréry, Rome, 1572. National Library of Sweden, Stockholm, KoB 1 ab.
61 Henricus Hondius, *Poli Arctici, et Circumiacentium Terrarum Description Novissima*, 1636. Barry Lawrence Ruderman Antique Maps Inc.
62 Alexander Keith Johnston, *Zoological Geography*, in *The Physical Atlas: a series of maps & illustrations of the geographical distribution of natural phenomena*, [Edinburgh], 1849. Oxford, Bodleian Library, Allen LRO 393.
63 Gerrit De Veer, *Curt van Nova Zembla*, in *Vraye description de trois voyages de mer ... faicts en trois ans ... par les navires d'Hollande et Zelande, au nord par derriere Norwege, Moscovie, et Tartarie*, Amsterdam, 1598. Oxford, Bodleian Library, Mason K 230, fol. 35r.
64 Gerrit De Veer, *Vraye description de trois voyages de mer ... faicts en trois ans ... par les navires d'Hollande et Zelande, au nord par derriere Norwege, Moscovie, et Tartarie*, Amsterdam, 1598. Oxford, Bodleian Library, Mason K 230, fol. 11v.
65 Gerrit De Veer, *Vraye description de trois voyages de mer ... faicts en trois ans ... par les navires d'Hollande et Zelande, au nord par derriere Norwege, Moscovie, et Tartarie*, Amsterdam, 1598. Oxford, Bodleian Library, Mason K 230, fol. 4r.
66 *Map of Franz Josef Land, showing journeys and discoveries of Frederick G. Jackson, F. R. G. S. Leader of the Jackson-Harmsworth Polar Expedition 1895–7*, Royal Geographical Society, London, 1898. London, Royal Geographical Society, 1898. Collection Allard Pierson UvA, HB-KZL 30.30.07; loan from the KNAG.
67 Harry Fisher, Sketch map of the lair of a polar bear. Scott Polar Research Institute, University of Cambridge, Harry Fisher collection, MS 287/28/2.
68 Frederick G. Jackson, *Infant contentment after dinner*, in *A Thousand Days in the Arctic*, Cambridge University Press, Cambridge, 1899. Oxford, Bodleian Library, 2035 d.18 (v.1), facing p. 176.

INDEX

Entries in italics refer to images

Adam of Bremen 31
Aleut International Association 10
Alexander the Great 23, 25–6
Arctic
    Athabaskan Council 10
    Circle 9
    cold 105, 114
    Council 10
    exploration 100, 124, 146–7, 169;
        see also individual explorers
    definition of 5, 9–10, 43, 164
    population of *124*, 127, 141, 144
Argo see Argonauts
Argonauts 84, *86*, 86–9
Arrowsmith, Aaron 121, *122–3*, 123–5
Atlantis 88

Baltic Sea 1, 41, 87, 89
Barents Sea 43
Barentsz, Willem 105, *167*, 166–7
Bartholomaeus Anglicus 146, *146*, 164
Behaim, Martin 93
Berghaus, Heinrich 164
Bering Strait 94–5, 130, *131*, 144, 161
Bering, Vitus 54
Blaeu, Willem *112–13*, 114
borders 34, 45, 121–3, 129, 141,
    158–9
boundary lines *see* borders
bow drill 130, 144, *144–5*, 146
Brendan of Clonfert *see* Saint Brendan
British Isles *see* England; Great Britain;
    Ireland
Bureus, Andreas *42*, 43, 45–6

Cagni, Umberto 55
Canada
    and fishing 149, 151–2
    expeditions to 124–5, 136–7, *137*, 140
    international collaboration 10
    north in 8
    on maps 117, *118*, 124–5, 134, *135*,
        148, *148*
Canary Islands 30, *71*, 74
cannibalism 82–3
Capell Brooke, Arthur de *44*, 45, 45–6,
    119, 121–2
Carta Marina
    and classical mythology 96, 110–11
    animals on 144, 156, 160, 165, 179
    motifs on 41, 75, *119*
    purpose of 1–2
cartes-à-figures 115, 116–21
cartouches 5, 114
Champlain, Samuel de *150–51*, 152–3
Chirikov, Aleksei 54
Clavus, Claudius 31, *32–3*, 34–5, 114
climate
    and indigenous land use 129, 134,
        135
    change 5, 9, 84, 129, 164, 174, 177–80
    north as cold 4, 9, 10, 16, 44, 46–7,
        50, 112–13, 114, 117, *118*, 147; *see
        also* climate, theory
    north as mild 43, 96–7, 153
    seasonal 41, 105, 153
    theory 37, 40–43, 46, 105
    see also frozen sea
codfish *see* fish

colouring of maps
    and power relations 64, 123–4, *158–9*
    impression of 43–6, 117, *158–9*
    luxury items 36
    polar bears 165
Columbus, Christopher 36, 51
Counter-Reformation 1–2, 78

Dass, Petter 155–6
decorative map borders *see cartes-à-figures*; cartouches
Delisle, Guillaume 74
deluge 60, *61*
Denmark
    colonial activities 110, 123, 128, 133
    international cooperation 10
    on maps 34, 34–5, 122–3
    religious allegiance 2
    see also Denmark Expedition
Denmark Expedition (1906–08) 128
Denmark–Norway *see* Denmark
De Wit, Frederick 115, *116–17*
'discoveries' *see* Arctic; knowledge; known unknown; North America
Don River 87
Dr Frankenstein's monster 100, 102
Droego 79, *80–81*, 84
Dulceti, Angelino 164
Dutch Republic
    and animals 105; *see also* Barentsz, Willem
    depictions of the north 112–13, 114, 160
    interests in the north 96–7
    see also cartes-à-figures

economic incentives 10, 50, 97, 117, 125, 127, 129–30, 145–6, 149, 152, 156, 161, 175, 179
Egypt 20, 22
encounters
    and mapping 106, 136–40
    between humans and animals 78, 144–5, 156, 165–7, *168*
    between people 21, 82, *111*, 130, *130–31*, 133
    with danger 84, 165
    with land 54, 82, *148*
    with mythical creatures 78, 107–8, 110
end of time, ideas about *see* eschatology
England
    on maps 27

    see also Great Britain
Engronelandt *see* Greenland
Eratosthenes 16–17
eschatology
    Christian 25–6, 108
    Islamic 22, 23
    Jewish 25
Estotiland 79, *80–81*, 81–2
eurocentrism 51
exotic north
    animals 78, 144, 145, 161, 165, 175
    climate 16, 47, 114 153
    cultural traits 2
    environment 102
    mythical creatures 96, 129
    peoples, ideas about 105–6, 110, 111, 114
expeditions, polar, *see* Jackson–Harmsworth expedition; Denmark Expedition; *see also entries on individual explorers*
exploration *see* expeditions; travel; *entries on individual explorers*
eyewitness accounts 89, 92, 97, 98
    lack of 20, 67
Ezekiel 25

Faḍlān, Aḥmad Ibn 21–2
Faroe Islands 70
Fastitocalon 78
Finland 10, 27, 87
fish
    as decorative elements 142, 144, 144–5
    as resource 119, 149–53, 179
    lack of 155–6, 179
    on maps 117, 118, 142, 150–51
    polar bear eating 165
Fisher, Harry 172, *173*
fishermen, experiences of 78, 82
France, interests in North America 150–51, 153, 155
Franklin, Sir John 9, 137–8, *139*, 140
Franz Josef Land 169, *170–71*, 172
frigid zone *see* climate theory
Frisland
    discussion of 79, 81–2, 84
    on maps 72–3, *80–81*, 109
Frobisher, Martin 97
frozen sea 11, 18, 28–9, 31, 41, 98
Fuca, Juan de 54
    strait of 124

210   MAPPING THE NORTH

geography
    historical 55, 60–64, 88–9
    imagined 4
    physical 161, *162–3*, 164
Germanus, Nicolaus 31, 34–5, *34–5*, 41
Germany
    globes from 93
    maps from *28–9*, 31, *34*, 36, 54–5,
        56–9, 164, *165*
Gilbert, Humphrey 96–7
global warming *see* climate change
globes
    impact on ideas of the north 93
    usage of 97
Gog and Magog
    as real peoples 22, 25–6
    on maps 26
    wall of *18–19*, 22, *23*, *24*, 108
    see also eschatology
Golden Fleece 84, 87
Great Britain
    explorers from 125, 136–40, *139*, 169
    ideas about north 5, 121–4, *122–3*
    promotion of 153–5, *154–5*
Greenland
    exploration of 55
    inhabitants of, ideas about 111, 114
    international cooperation 10
    location *32–3*, *34–5*, 36, 82
    see also Denmark, colonial activities;
        Denmark expedition; Norse,
        settlements on Greenland; Peary,
        mapping of Greenland
Gwich'in Council International 10

Hearne, Samuel 125
Hereford *mappa mundi* 71, 74, 107–8, 110
Herodotos 111
Higden, Ranulf 27, 30
Holl, Leinhart 31, 36
Holm, Gustav 133
Homann Johann Baptist 160
Hondius, Henricus *158–9*, *160–61*
Hornius, Georgius *86*, 89
Hudson Bay 54, 117, *118*, 136
Hyperboreans 107

Icaria 79, *80–81*
ice
    animals and 174
    as a hindrance to explorers 97, 106,
        138
    on maps 5, *38–9*, 41, 46, *165*
    travels on 130, *130–31*, 134, *135*
    see also frozen sea
ice floes *see* ice, on maps
Iceland
    and fishing 152
    animals in 164, 179
    identification with Thule 18, 79
    international collaboration 10
    on maps 27, *32–3*, 35–6, *38–9*
    volcanic activity 84
Icelandic Sagas *see* Norse, literature
identity 3, 5, 10, 40, 105, 108–10; *see also*
    exotic north; patriotism
Ikmallik *137*, 137–8
Iligliuk 134, *135*, 136, *137*
immram 70
indigenous *see* Inuit; Sami
Insulae Fortunatae *see* Canary Islands
Inuit
    animal relations 144
    art 144
    Circumpolar Council 10
    knowledge 137–8, *139*, 140, 147
    maps 134, *135*, *137*
    populations 124, *124*, 129
Inventio Fortunata 92–3, 97–8, 100
Ireland
    home of Saint Brendan 69
    on maps 27, 30
Iskandar *see* Alexander the Great
Islamic mapping *18–19*, 20–22

Jackson, Fredrick 169, 172
Jackson–Harmsworth expedition 169,
    170–71, 172, *173*, *174*
Jefferys, Thomas 153, *154–5*
Johnston, Alexander Keith *162–3*
journeys *see* travel

Kamchatka Peninsula 54
kayak 119, *119*, 130, *130–31*, 134, *135*
King William Island *139*, 140
Kircher, Athanasius 98, *99*, *100–101*,
    102
knowledge
    circulation of 2, 82, 92, 106
    conceptions of authoritative 67–8,
        89, 97, 98, 102, 103, 120; *see also*
        Ptolemy, outdated
    first-hand *see* eyewitness accounts
known unknown 13, 16–17, 31, *48–9*,
    50–51, 60, 65, 98, 105
Kraken 78

Kuniit 132–4, 141
Kyiv 22

Lapland *see* Sapmi
Leibniz, Gottfried Wilhelm 89
Little Ice Age 41
Lofoten 93

McClintock, Sir Francis Leopold 138
Macrobius 37, 40
maelstrom *86*, 93, 96
Magnus, Olaus
   History of the Nordic Peoples 111, 121
   see also Carta Marina
Magog *see* Gog and Magog
Malangen fjord 45
mappa mundi
   definition of 26, 74
   images on 107–8, 114
   see also Hereford *mappa mundi*;
     Higden, Ranulf; Psalter map
Mediterranean Sea, and mapping 17, 22, 26, 87–8
Mercator, Gerhard 92–3, *94–5*, 96
Mikkelsen, Ejnar 128
Minard, Charles-Joseph *47*, 47, 50
monsters *see* mythical creatures
Moskstraumen *86*, 93, *94–5*, 96
Moucheron, Balthasar 96
Mount Ararat 60
Müller, Gerhard Friedrich 51, *52–3*, 54
mythical creatures 26, 107, 165
   in the sea 43, 75, 78, 96, 143, 156, 157, 169
   see also sciapods
mythical islands *see* Droego; Estotiland; Frisland; Icaria; Saint Brendan
mythology, Greco-Roman 43, 84–9

naming 36, 54, 74, *94–5*, *122–3*, 125–9, 141, 149, 160, 172
Nansen, Fridtjof 55, 106
Napoleon's Russian campaign *47*, 50
Netherlands *see* Dutch Republic
Newfoundland 52, 149, 153
New World *see* America
Nordic region
   animals in 144, 165
   climate 43, 45–6
   knowledge of 2, 17, 34–5, 152, 156
   map of 31, *32–3*, *34–5*
   mythical 108, 110, 114
   Sami population *see* Sami

Norse
   in the east 21
   legacy of *109*, 110, 133
   literature 108
   settlements on Greenland 108, 110
North America
   animals in 165
   exploration of 51, 54, *62–3*, 78, 79, 82–3, 148–52; *see also* entries on individual explorers
   mapping of 149, 152
   Norse in 108, *109*
   portrayal of 96–7, 123, 153
   scenes from 117, 118
Northeast Passage 36, 54, 96–7, 167
North Pole
   expeditions to 3, 55, 92, 125, 147
   inhabitants 92–3, 114
   maps of 54, 56–7, 58–9, 93, 117, 121–3, *122–3*
   open sea at 92, 96–8, 100–102
Northwest Passage 36, 54, 96–7, 136–8, 149
Norway
   fishing 152, 155–6
   identification of 18
   international collaboration 10
   on maps 34, *34–5*, 44, 46
   territory of 45, 127
Novaya Zemlya 105, *166–7*, 167

ocean floors, mapping of 179
Ortelius, Abraham
   depictions of ice *38–9*, 41, *48–9*, 153
   geographical knowledge 50–51, 79, 83
   historical mapping see *Parergon*
   map of northern Europe *72–3*, 114
   mythical islands *72–3*, 74, 79, 81–2
   polar bears 41, 165

Paradise
   on maps 30
   searching for 69–70
parchment 146
Parergon 88, *90–1*
Parry, Sir William 136–7
patriotism 41, 46, 110
Peary, Robert E.
   and the mapping of Greenland 55, 127–9
   claim of reaching the North Pole 3, 55

conceptions of exploration and
    mapping 3, 146–7
  in winter dress 147
Perthes, Justus 54, 56–7, 58–9
Pilappelandt 36
Plato 85, 88
Plautius, Caspar 76–7, 78, 82
Pliny the Elder 111
polar bears
  and climate change 174, 178
  as curiosities 105, 165, 174
  as danger 165–9
  as object of study 162–3, 172, 173
  as resource 147, 169
  cubs 172, 173, 174, 174
  name 164
  on maps 38–9, 41, 46, 162–3
Polar Circle, northern see Arctic Circle
politics
  and ideas about the north 5, 8–9, 65, 178
  and mapping 2, 45, 64, 78
  claims over territories 45, 110, 123–4, 128–30
Psalter map 24
Ptolemy, Claudius
  knowledge of the north 16–18
  in medieval Islamic mapping 20–21
  in the Renaissance 30–31, 85
  outdated 21, 34
'pygmies', on maps of the north 111, 111
Pytheas of Massalia 17–18, 20

Quin, Edward 60, 61, 62–3, 64

Raleigh, Walter 97
Reformation 1–2
religion and maps 16, 22, 24, 26, 30, 34, 60, 74, 98–100; see also Counter-Reformation; Islamic mapping; Reformation
Ribero, Diego 148, 148, 152, 153
Ross, James C. 137, 137
Rudbeck, Olof the Elder
  and mapping 86, 87–9
  biography 87–8
  evaluation of sources 89, 97, 102
  opinions about 89, 103
  portrait of 85
Rūsiyyah 21
Russia
  animals in 165, 167
  expansion east 51, 52–3, 54
  hardships of war 50
  location of 8, 123
  trade with 118
Russian Association of Indigenous
  Peoples of the North 10
Ruysch, Johannes 93

Saint Brendan
  biography 69
  on maps 71, 72–3, 74, 103
  purpose of account 70, 79
Sami
  Council 110
  means of transportation 43, 45, 119
  stereotypes of 43, 119, 119, 120, 121
Sapmi 118–19, 129, 158–9
Saxo Grammaticus 31
Scandinavia see Nordic region
Schefferus, Johannes 119, 120, 119–21
sciapods 107–8, 110, 129
Scoresby Jr, William 125, 126, 127, 129
Seven Years War 153, 154–5, 155
Shelley, Mary 100, 102
Shetland Islands 18
skiing 41–5, 42
Sorant 81
South Pole 93, 98, 99, 125
Soviet Union 8
Spitsbergen see Svalbard
Stielers Hand-Atlas 54–5, 56–9, 125
Strait of Anian see Bering Strait
Svalbard 125, 126, 127
Sweden
  borders of 45
  international cooperation 10
  on maps 6–7, 34–5, 44, 46
  promotion of 1, 6–7, 8, 41, 43, 46, 85, 88
  religious allegiance 2

Thule 16–20
  location of 16, 18–20, 79
  on maps 14–15, 21, 27
  origin 17–18
trade see economic incentives
travel
  and mapping 3, 67–9
  armchair 46, 178
  fictional 67–9, 83, 84, 89, 92
  hardships of 47, 50, 69, 108, 147
  religious 32–3, 34, 70, 75
  routes on maps 134, 135
  winter 45

INDEX   213

travelogue *see* travel writing
travel writers *see* Aḥmad Ibn Faḍlān; Arthur de Capell Brooke; Samuel de Champlain; Pytheas of Massalia; Saint Brendan; Nicolò Zen
travel writing
   genre 67–9
   religious purpose 70, 78
   used by mapmakers 79, 88–9, 102–3, 128, 141, 178

Ultima Thule 20
USA
   international collaboration 10
   promotion of 128, 154–5
   see also North America
'us' and 'them' *see* identity; exotic north

whale
   as an island 67, 69–70, 75, *75*, 76–7, 78
   as commodity 119, *119*, 145, 161, *159–60*, 164
   exotic 79, 156, *157*, 161
   fishing 125, 127, 144
   study of *162–3*, 164

Veer, Gerrit de 166–7, *167*, *168*, 169
Verelius, Olof 88
Vikings *see* Norse
Vinland Sagas see Norse, literature
Volga River 21, 87

Waldsemüller, Martin 165
weather *see* climate
world maps
   early modern European 48–9, *50*, 92–3, 148, 165
   Islamic *18–19*, 20–22
   medieval European
      see mappa mundi
   Ptolemaic *14–5*, 17, *28–9*, *30–31*
   shaping world-views 141

Zealand 35
Zen, Antonio 81–3
Zen, Nicolò the Elder 81–3
Zen, Nicolò the Younger *80–81*, 81–3
Zichmni, Prince 81–2, 84